73

Advances in Biochemical Engineering/Biotechnology

Managing Editor: T. Scheper

Springer-Verlag
Berlin Heidelberg GmbH

Metabolic Engineering

Volume Editor: Jens Nielsen

With contributions by
L. Eggeling, J. Dynesen, M. Gárdonyi, R. T. Gill,
A. A. de Graaf, B. Hahn-Hägerdal, L. J. Jönsson,
C. Khosla, P. Licari, R. McDaniel, M. McIntyre,
C. Müller, J. Nielsen, R. R. Cordero Otero, H. Sahm,
U. Sauer, D. E. Stafford, G. Stephanopoulos,
C. F. Wahlbom, K. S. Yanagimachi, W. H. van Zyl

Springer

Advances in Biochemical Engineering/Biotechnology reviews actual trends in modern biotechnology. Its aim is to cover all aspects of this interdisciplinary technology where knowledge, methods and expertise are required for chemistry, biochemistry, microbiology, genetics, chemical engineering and computer science. Special volumes are dedicated to selected topics which focus on new biotechnological products and new processes for their synthesis and purification. They give the state-of-the-art of a topic in a comprehensive way thus being a valuable source for the next 3–5 years. It also discusses new discoveries and applications.

In general, special volumes are edited by well known guest editors. The managing editor and publisher will however always be pleased to receive suggestions and supplementary information. Manuscripts are accepted in English.

In references Advances in Biochemical Engineering/Biotechnology is abbreviated as *Adv Biochem Engin/Biotechnol* as a journal.
Visit the ABE home page at http://link.springer.de/series/abe/
http://link.springer-ny.com/series/abe/

ISSN 0724-6145
ISBN 978-3-642-07534-6 ISBN 978-3-540-45300-0 (eBook)
DOI 10.1007/978-3-540-45300-0

Library of Congress Catalog Card Number 72-152360

Typesetting: Fotosatz-Service Köhler GmbH, Würzburg
Cover: E. Kirchner, Heidelberg
Printed on acid-free paper SPIN: 10792316 02/3020hu - 5 4 3 2 1 0

Managing Editor

Prof. Dr. T. Scheper
Institute of Technical Chemistry
University of Hannover
Callinstraße 3
30167 Hannover, Germany
E-mail: scheper@mbox.iftc.uni-hannover.de

Volume Editor

Prof. Dr. J. Nielsen
Center for Process Biotechnology
Department of Biotechnology
Building 223
Technical University of Denmark
2800 Lyngby
Denmark
E-mail: jn@ibt.dtu.dk

Editorial Board

Advances in Biochemical Engineering/Biotechnology Now Also Available Electronically

For all customers with a standing order for Advances in Biochemical Engineering/Biotechnology we offer the electronic form via LINK free of charge. Please contact your librarian who can receive a password for free access to the full articles. By registration at:

http://www.springer.de/series/abe/reg_form.htm

If you do not have a standing order you can nevertheless browse through the table of contents of the volumes and the abstracts of each article at:

http://link.springer.de/series/abe/
http://link.springer_ny.com/series/abe/

There you will find also information about the

- Editorial Bord
- Aims and Scope
- Instructions for Authors

Attention all Users of the Springer Handbook of Enzymes

Information on this handbook can be found on the internet at
http://www.springer.de/enzymes/

A complete list of all enzyme entries either as an alphabetical Name Index or as the EC-Number Index is available at the above mentioned URL. You can download and print them free of charge.

A complete list of all synonyms (more than 20,000 entries) used for the enyzmes is available in print form, ISBN 3-540-41830-X.

Save 15%

We recommend a standing order for the series to ensure you automatically receive all volumes and all supplements and save 15% on the list price.

Preface

With the introduction of genetic engineering of *Escherichia coli* by Cohen, Boyer and co-workers in 1973, the way was paved for a completely new approach to optimisation of existing biotech processes and development of completely new ones. This lead to new biotech processes for the production of recombinant proteins, e.g. the production of human insulin by a recombinant *E. coli*. With the further development in genetic engineering techniques the possibility ofto applying this for optimisation of classical fermentation processes soon became obvious, and advancements in genetic engineering allowed a far more rational approach to strain improvement than the classical approach of mutagenesis and screening, namely introduction of directed genetic changes through rDNA technology. In 1991, this led Bailey to discuss the emerging of a new science called metabolic engineering, which he defined as "*the improvement of cellular activities by manipulations of enzymatic, transport, and regulatory functions of the cell with the use of recombinant DNA technology*". Initially metabolic engineering was simply the technological manifestation of applied molecular biology, but with the rapid development in new analytical- and cloning techniques, it has become possible to introduce directed genetic changes rapidly and subsequently analyse the consequences of the introduced changes at the cellular level.

In recent years, there has been a rapid development in the field of metabolic engineering, and this has resulted in extensive number of reviews in the field (see e.g. Nielsen, 2001)., There has been one text book describing the principles and methodologies of metabolic engineering (Stephanopoulos et al., 1998), and a multi-author book with many excellent examples of metabolic engineering edited by Lee and Papoutsakis (1999). A journal fully devoted to this topic has appeared (www.apnet.com/mbe), there are sessions on metabolic engineering at most conferences on biochemical engineering and applied microbiology, and a conference series devoted to this topic has developed. With this extensive coverage of this rapidly growing research field, it is impossible to cover all aspects of metabolic engineering in a single issue of Advances in Biochemical Engineering/Biotechnology. However, several key examples of metabolic engineering will be reviewed in this volume:

- Improvement of yield and productivity – exemplified by amino acid production by *Corynebacterium*
- Production of novel compounds – exemplified by the overproduction of novel polyketides

- Extension of substrate range – exemplified by engineering of *Saccharomyces cerevisiae* for xylose utilisation
- Development of novel biosynthetic routes that may replace chemical synthesis routes – exemplified by engineering of indene bioconversion
- Improvement of cellular properties – exemplified by engineering of the morphology of *Aspergillus*

In addition, new concepts for selection of strains with improved properties are discussed – here referred to as evolutionary engineering. Finally, Stephanopoulos and Gill discuss the status of Metabolic Engineering, and predicts an expanded role for this field in the future.

I hope that you will enjoy reading the chapters.

Spring 2001 Jens Nielsen

References

Bailey JE (1991) Science 252:1668—1674
Lee SY, Papoutsakis ET (1999) Metabolic engineering. Marcel Dekker, New York
Nielsen (2001) Appl Microbiol Biotechnol (in press)
Stephanopoulos G, Aristodou A, Nielsen J (1998) Metabolic engineering. Academic Press, San Diego

Contents

After a Decade of Progress, an Expanded Role for Metabolic Engineering
G. Stephanopoulos, R. T. Gill 1

Metabolic Engineering for L-Lysine Production by *Corynebacterium glutamicum*
A. A. de Graaf, L. Eggeling, H. Sahm 9

Process Development and Metabolic Engineering for the Overproduction of Natural and Unnatural Polyketides
R. McDaniel, P. Licari, C. Khosla 31

Metabolic Engineering of *Saccharomyces cerevisiae* for Xylose Utilization
B. Hahn-Hägerdal, C. F. Wahlbom, M. Gárdonyi, W. H. van Zyl,
R. R. Cordero Otero, L. J. Jönsson 53

Metabolic Engineering of Indene Bioconversion in *Rhodococcus* sp.
D. E. Stafford, K. S. Yanagimachi, G. Stephanopoulos 85

Metabolic Engineering of the Morphology of *Aspergillus*
M. McIntyre, C. Müller, J. Dynesen, J. Nielsen 103

Evolutionary Engineering of Industrially Important Microbial Phenotypes
U. Sauer 129

Author Index Volumes 51–73 171

Subject Index 181

After a Decade of Progress, an Expanded Role for Metabolic Engineering

Gregory Stephanopoulos, Ryan T. Gill

Department of Chemical Engineering, MIT Room 56-469, Cambridge MA 02139, USA,
e-mail: gregstep@mit.edu

Over the past decade, metabolic engineering has emerged as an active and distinct discipline characterized by its over-arching emphasis on *integration*. In practice, metabolic engineering is the directed improvement of cellular properties through the application of modern genetic methods. Although it was applied on an ad hoc basis for several years following the introduction of recombinant techniques [1, 2], metabolic engineering was formally defined as a new field approximately a decade ago [3]. Since that time, many creative applications, directed primarily to metabolite overproduction, have been reported [4]. In parallel, recent advances in the resolution and acquisition time of biological data, especially structural and functional genomics, has amplified interest in the systemic view of biology that metabolic engineering provides. To facilitate the burgeoning scientific exchange in this area on a more regular and convenient basis, a new conference series was launched in 1996 followed by a new journal in 1999.

Keywords. MetabolicEngineering, Functional, Genomics, Phenotype, Systems, Biology

1 Introduction 1

2 Expanded Spectrum of Applications for Metabolic Engineering . . 2

3 New Technologies for Probing the Cellular Phenotype 5

4 Metabolic Engineering and Functional Genomics 6

5 Closure 7

References 8

1 Introduction

Metabolic engineering is distinguished from previous ad hoc genetic strategies by a step of analysis whereby the physiological impact of the genetic modifications carried out is rigorously assessed. As a result, the next round of genetic manipulations is performed in a directed rather than random manner. This iterative approach to cell improvement constituted a significant departure from prior practice dominated by single gene overexpression. Moreover, it reflected an increasing appreciation of the fact that control of metabolite synthesis does not reside in a single rate-limiting step. Rather, control is distributed among several reaction steps in a pathway, as suggested previously by the pioneers of Metabolic Control Analysis [5–9]. An important consequence of this realization was the need for a more detailed evaluation of the cellular physiological state that goes beyond the macroscopic evaluation of metabolite uptake and production rates. Enumeration and quantification of intracellular metabolic fluxes provided this additional information and Metabolic Flux Analysis [10, 11] emerged as a distinctive focus of metabolic engineering. Another distinguishing feature of metabolic engineering is its emphasis on *integration*. This was pointed out by drawing attention to the properties of metabolic networks in their entirety in contrast to the prior focus on single reactions in a pathway. The flux is the most important property of a metabolic network. Fluxes are systemic network properties and the development of methods to measure fluxes and understand control of flux is a key objective of metabolic engineering.

To recap, metabolic engineering is about pathway modification at the genetic level and evaluation of the ensuing cellular physiology. It also concerns itself with the systemic properties of metabolic networks and in particular metabolic flux and its control. This paradigm has proven very fruitful in general cell improvement including enhanced product yield and productivity [12–14], an expanded range of substrate utilization [15–17], formation of novel products [18–21], and improved cellular properties [22, 23].

In view of this progress, what is the outlook for the next decade? First, metabolic engineering will continue along the very successful path of the past, producing more fascinating examples of cell improvement in diverse areas of biotechnology. Second, metabolic engineering has a unique opportunity to expand its role by virtue of its strong focus on *integration* and the incorporation of new experimental and computational tools. As such, metabolic engineering provides a convenient framework that can accommodate the massive movement of biological sciences towards *experimentally* based, system-wide analysis. A further application of metabolic engineering principles will be in the design of system wide experiments, i. e., what experiments should be run to allow maximum evaluation of the regulatory network under study. Also, the role of metabolic engineers in the process of biological discovery will expand as new technologies continue to increase the size and resolution of regulatory databases. The above assertions are supported by the genomics revolution, an ever-expanding infrastructure of applied molecular biology, and numerous emerging applications of biotechnology in the production of chemicals and materials, as well as in the medical field. These possibilities suggest an expanded role for metabolic engineering, as outlined below, due to a broader spectrum of applications, new powerful tools for studying cell physiology, and a direct involvement in the field of functional genomics.

2 Expanded Spectrum of Applications for Metabolic Engineering

While in the past metabolic engineering focused primarily on enhancing strain productivity, expanding substrate utilization range, and forming novel products, the future spectrum of applications for metabolic engineering has expanded hand in hand with the explosive growth in biological research. Driving this expanded role has been the massive efforts towards evaluating system-wide biological properties. For example, the full sequence of 42 organisms is currently complete with an additional 250 organisms in process (http://ncbi.nlm.nih.gov). Functional genomic technologies are also in place that allow the activity of complete genomes to be observed, proteomic techniques are increasingly being demonstrated, and improved methods of measuring metabolic fluxes are developing rapidly. As a result of these developments, we envision three primary areas of research that an expanded metabolic engineering will impact greatly. First, traditional metabolite overproduction will benefit as global regulatory data accumulate and the effects of directed alterations are resolved at much greater physiological detail. Second, the spectrum of alternative host organisms and relevant gene products will continue to expand as full genomes of plants, fungi, bacteria, and mammals are sequenced. Finally, biocatalytic applications for the production of chiral molecules will progress as we begin to understand the systemic properties that favor the production of stereospecific compounds. Importantly, developments in each of these research areas will be mutually beneficial. That is, the expanded host and gene product range will enhance the production of chiral molecules.

Although most applications of the past decade and obvious future extensions focus on the improvement of industrial strains for metabolite overproduction, perhaps an even greater impact of metabolic engineering will be in genetic therapy, pharmaceutical diagnostic assays, or programs of drug discovery. Although issues of delivery presently dominate the prospects of gene therapy, the ultimate success of this very promising approach will depend on the correct identification of the target(s) of genetic intervention. As such, the central problem of gene therapy will be no different to that of strain improvement and a systemic analysis of genomic and physiological measurements will play an important role in this area. Moreover, assessing the specific physiological phenotypes observed after overexpression of specific gene therapeutics is an obvious extension of more traditional metabolic engineering systems.

Another unconventional application of metabolic engineering is the development of targets for the screening of compound libraries in drug discovery. The key concept here is that single enzyme assays are becoming less effective in identifying robust lead molecules with high probability of maintaining activity under in vivo conditions, for the simple reason that it is less likely that a single enzyme is responsible for most systemic diseases [24]. This means that drugs effective against more than one target will have a higher probability of success and fewer side effects. Additionally, identification of lead molecules will have to rely increasingly on the response of multiple markers of cellular function as opposed to a single marker-based selection that is presently the norm. The above characteristics constitute drastic departure from current practice in drug discovery, yet they are entirely within the realm of feasibility given a suitable intellectual framework and sufficient measurements about the cellular state. Such a framework of integration is available from metabolic engineering whose power will be further enhanced with the inclusion of the new methods for probing the cellular phenotype.

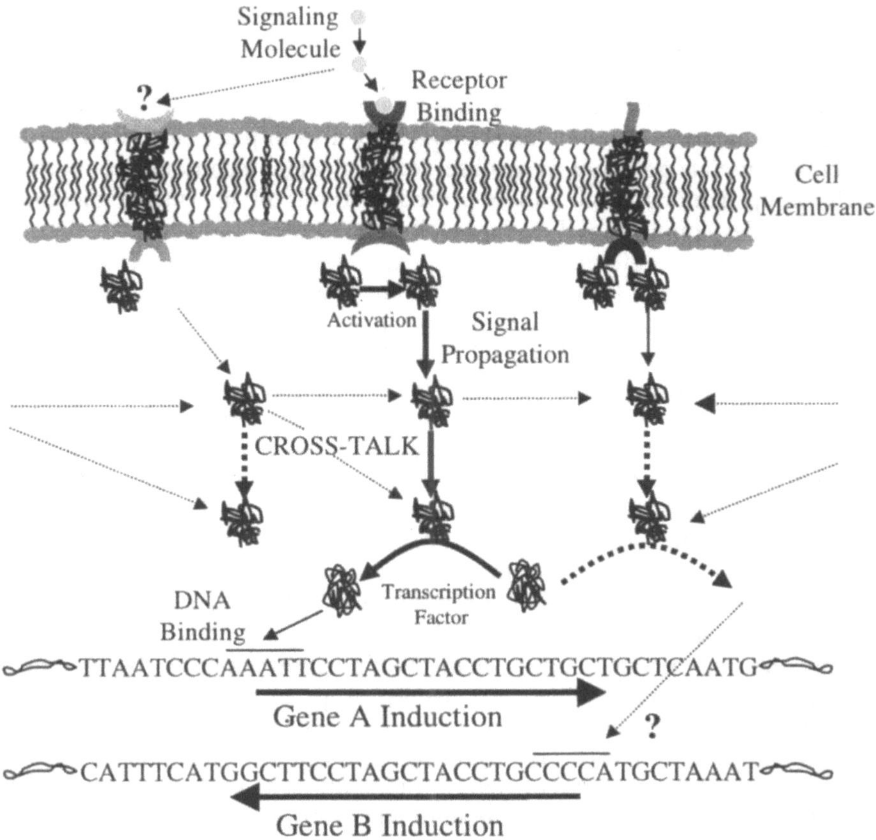

Fig. 1. Representation of signal transduction pathways. Signaling molecules bind to receptor proteins on the outside of the cell membrane. The receptor protein is activated (typically by conformational changes) on the interior side of the cell membrane. The activated protein next transfers an interior signaling molecule to a second signal transduction protein, followed by a third, etc. The end result is the activation of a DNA binding protein, a transcription factor, transcription initiation, and gene induction. Cross-talk occurs when signaling molecules are transferred across signaling pathways leading to the activation of different transcription factors and ultimately inducing different genes. Also, non-specific binding of extra-cellular signaling molecules can lead to partial activation of alternative signaling pathways

A final non-obvious but very important future role for metabolic engineering will be the analysis of signal transduction pathways. Signal transduction pathways are involved in inter-cellular interactions and communication of extra-cellular conditions to the interior of the cell. Signaling occurs via consecutive phosphorylation-dephosphorylation steps whereby the phosphorylated (active) form of an intermediate protein acts as a catalyst (kinase) for the phosphorylation of the subsequent step. The final outcome of a signaling pathway is often the activation of a transcription factor that, in turn, initiates gene expression [25]. To date, signal transduction pathways have been investigated in isolation from one another. It has become abundantly clear, however, that there is a great degree of interaction (cross-talk) of signal transduction pathways for the simple reason that they share common protein intermediates [26]. This introduces the possibility that one ligand may effect the expression of more than one gene or that the expression of a single gene may be effected by more than one ligand (Fig. 1). Again, the network features of signaling provide a fertile ground for the application of concepts from metabolic engineering in conjunction with expression and, in particular, proteomics data. Certain modifications influence to a significant extent gene expression and, as such, will have to be made to account for the fact that signaling pathways catalyze the propagation of information compared to interconversion of molecular species characterizing metabolic pathways. The correct formulation and applicable principles that take this difference into consideration are yet to be developed.

3
New Technologies for Probing the Cellular Phenotype

DNA micro-arrays are the basis of powerful new technologies for the simultaneous measurement of the amount of specific DNA sequences in a heterogeneous mixture of hundreds of thousands of nucleic acids (cDNA, RNA, DNA) [27]. The basis for DNA micro-array studies is the tendency of complementary nucleic acid strands to form stable, double stranded hybrids. The stability of these hybrids decreases as the number of perfectly matched nucleotides decreases, as well as at high temperatures or in the absence of sufficient buffering capacity. By covalently binding fluorescent nucleotides to the target nucleic acid sample and hybridizing to the micro-array of DNA probes, complementary DNA strands will associate and fluoresce. The intensity of the fluorescent signal from each DNA probe on a micro-array is indicative of the amount of complementary DNA in the target solution. As a result of the availability of numerous fluorescent molecules, several DNA target solutions can be probed in parallel on the same micro-array. Fluorescent intensity ratios from each DNA probe then reflect the relative amount of complementary DNA in each target solution. Using this technology, expression levels for up to 30,000 genes have been measured in parallel (http://www.tigr.org). Prototype oligonucleotide micro-arrays currently contain up to 800,000 features with higher density arrays still in development (personal communication). Recent total size estimates for the human genome range between 40,000 genes and 130,000 genes, a range easily contained on soon-to-be-available micro-arrays. Thus, future studies of full genome transcriptional regulation for any organism of biotechnological relevance are imminent realities. Importantly, many of the developments in functional genomic studies have directly enhanced the development of proteomic technologies. For example, antibody based micro-arrays can be synthesized, imaged, quantified, and evaluated using DNA micro-array techniques. In addition, enhanced two-dimensional gel electrophoresis methods and integrated peptide analysis by LC-MS are in development. Although not at the same level as DNA micro-array studies, the importance and activity in proteomics suggests that developments in this area will accelerate in the near future. Given the similar forms of current genomic and future proteomic data sets, an established analytical framework from functional genomics should be directly applicable to proteomic studies.

To understand, however, cellular function and the correlation between gene expression and the actual physiological state of the cell, we need to be able to determine the latter with high accuracy. How the physiological state of the cell is defined ultimately will determine the utility of gene expression data. That is, enzymatic activity is a function of not only the associated mRNA concentration but also the enzyme concentration, cofactors, antagonist molecules, pH, redox potential, proper folding, proteases, and scores of additional cellular features which help to define the physiologic state of the cell. The set of intracellular fluxes represents the interaction of all of these features; namely, the actual rate at which metabolites are processed throughout the metabolic network is the outcome of all of the aforementioned variables and most directly reflects the physiological state of the cell. Therefore, the set of technologies probing the intracellular make up and function needs to be complemented with methods of commensurate resolution in determining intracellular metabolic fluxes as measures of cell physiology and function. Flux determination has been carried out to date by extra-cellular metabolite measurements combined with metabolite balances. Occasionally, stable isotopic tracers have also been used to produce flux estimates of previously unobservable fluxes. Clearly, we need to expand the number of fluxes that can be reliably observed to allow a more direct comparison with the available data of the expression phenotype. An exciting new approach to expanding the range of metabolic flux measurements relies upon the use of gas chromatography-mass spectrophotometry (GC-MS) and nuclear magnetic resonance (NMR) [28, 29]. An analytical framework has been established and experimental techniques are rapidly developing that allow for enumerating complete isotopomer balances and solving for isotopomer content as a function of metabolic flux. For example, Pedersen et al. [29] recently utilized this GC-MS-based approach to characterize an oxalic acid non-producing strain of *Aspergillus niger*. Fluxes so determined are robust in that they satisfy a great degree of redundancy and thus are extremely sensitive to variations of the intracellular state. These are only three of the technologies that we believe will expand the scope of future metabolic engineering studies.

4
Metabolic Engineering and Functional Genomics

Besides assigning function to (annotating) newly sequenced open reading frames (ORFs), another goal of functional genomics is to integrate genomic, expression, and proteomic data in order to produce a more comprehensive picture of the cellular functions. This objective, of course, is very similar to the central theme of metabolic engineering of elucidating the architecture of cellular control as an integral part of the directed cellular improvement process. As such, there is substantial synergism and a strong bi-directional relationship between the goals and tools of metabolic engineering and functional genomics. First, metabolic engineering provides an integrated, system theoretic framework for analyzing the data generated from the above technologies. At the same time, metabolic engineering can benefit immensely from the information that will be extracted from such data. Think, for a moment, of identifying the expression profiles associated with high productivity periods in the course of a fermentation. Or, similarly, isolating a set of differentiating genes and their characteristic expression pattern that are associated with the onset of a particular disease, especially the dynamic sequence of expression profiles as the disease evolves with time. Importantly, a specific outcome of functional genomic studies is genes whose expression patterns are indicative of particular physiological states. Therefore, micro-arrays can be viewed as ultra-high dimensional biosensors with many far-reaching applications. As methods improve for obtaining expression data on- or off-line within minutes, the need of appropriate indicator genes or proteins will grow. It would be unfortunate, however, to restrict DNA micro-arrays to roles of biosensors. With a conscious effort towards the consilience of metabolic engineering principles and functional genomic data and desires both fields will benefit and progress rapidly. The previously mentioned examples and the clear overlap between these fields fuel the growing excitement about genomic and other derivative technologies and the implications for biomedical research in general.

5
Closure

Biological research is witnessing a return to the systems view of biology [30] with the advent of several technologies that provide such data. As a result, we foresee a new decade of great progress for metabolic engineering. There are, however, several problems to overcome in realizing the potential previously described. In contrast to the impressive progress in the development of methods and instrumentation for probing the intracellular state and function, systematic methods for the effective analysis of such data have received rather scant attention. Data evaluation is usually limited to cursory inspections by the user or, at best, to automated spot comparison (spot-oriented analysis) and rudimentary statistical analysis. Furthermore, faced with information overload, there is a natural tendency to focus subjectively on what is viewed a priori as relevant or important and relegate everything else to the background. Most importantly, besides methods and algorithms, there is a scarcity of experienced personnel who have the computational skills to develop such technologies and use them for extracting important information from the above data sets. These limitations are receiving broad attention presently calling for innovative approaches to provide much needed solutions.

Metabolic engineering with its focus on integration provides an appropriate framework for analyzing system-wide databases as well as for the design of experiments that maximize the useful information that can be extracted from them. The marrying of synthesis and analysis steps is a core feature of metabolic engineering and, as a result, an expanded role for metabolic engineering is anticipated. Given all of the above opportunities, we envision metabolic engineering principles as the basis, a starting point, for future systemic studies.

These principles will be applied in the design of systemic studies of not only strain improvement or metabolite overproduction but also in functional genomics, signal transduction, drug discovery, and gene therapy, among others. The value of a consensus theoretical framework will be realized through enhanced communication and collaboration with benefits for bioprocess engineering as well as biological discovery and medical research in general.

References

1. Saiki R, Scharf S, Faloona F, Mullis K, Horn G, Erlich H, Arnheim N (1985) Science 230:1350–1354
2. Goeddel D, Kleid D, Bolivar F, Heyneker H, Yansura D, Crea R, Hirose T, Kraszewski A, Itakura K, Riggs A (1979) Proc Nat Acad Sci 76:106–110
3. Bailey J (1991) Science 252:1668–1674
4. Stephanopoulos G, Aristidou A, Nielsen J (1998) Metabolic Engineering. Academic Press
5. Savageau MA (1970) J Theor Biol 26:215–226
6. Savageau MA (1969) J Theor Biol 25:370–379
7. Savageau MA (1969) J Theor Biol 25:365–369
8. Kacser H, Burns J (1973) Symp Soc Exp Biol 27:65–104
9. Heinrich R, Rapoport TA (1974) Eur J Biochem 42:89–95
10. Stephanopoulos G, Vallino J (1991) Science 252:1675–1681
11. Stephanopoulos G (1998) Biotechnol Bioeng 58:119–120
12. Kinoshita S, Udaka S, Shimono M (1957) J Gen Appl Microbiol 3:193–205
13. Ingram L, Conway T, Clark D, Sewell G, Preston J (1987) Appl Environ Microbiol 53:2420–2425
14. Colon G, Jetten M, Nguyen T, Gubler M, Follettie M, Sinskey A, Stephanopoulos G (1995) Appl Environ Microbiol 61:74–78
15. Burchhardt G, Ingram L (1992) Appl Environ Microbiol 58:1128–1133
16. Carey V, Walia S, Ingram, L (1983) Appl Environ Microbiol 46:1163–1168
17. Lawlis V, Dennis M, Chen E, Smith D, Henner D (1984) Appl Environ Microbiol 47:15–21
18. Anderson S, Berman-Marks C, Lazarus R, Miller J, Stafford K, Seymour J, Light D, Rastetter W, Estell D (1985) Science 230:144–148
19. McDaniel R, Ebert-Khosla S, Hopwood D, Khosla C (1993) Science 262:1546–1550
20. Peoples O, Liebl W, Bodis M, Maeng P, Follettie M, Archer J, Sinskey A (1988) Mol Microbiol 2:63–72
21. Sabatie J, Speck D, Reymund J, Hebert C, Caussin L, Weltin D, Gloeckler R, O'Regan M, Bernard S, Ledoux C, Oshawa I, Kamogawa K, Lemoine Y, Brown S (1991) J Biotechnol 20:29–50
22. Windass J, Worsey M, Pioli E, Pioli D, Barth P, Atherton K, Dart E, Byrom D, Powell K, Senior P (1980) Nature 287:396–400
23. Khosla C, Bailey J (1988) Nature 331:633–634
24. Kelsoe JR, Ginns EI, Egeland J, Gerhard D, Goldstein A, Bale S, Pauls D, Kidd K, Conte G et al. (1989) Nature 342:238–243
25. Lauffenburger D, Linderman J (1996) Receptors: models for binding, traficking, and signaling. Oxford University Press
26. Roberts C, Nelson B, Marton M, Stoughton R, Meyer M, Bennett H, He Y, Dai H, Walker W, Hughes T, Tyers M, Boone C, Friend S (2000) Science 287:873–880
27. Fodor S, Leighton J, Pirrung C, Stryer L, Lu A, Solas D (1991) Science 251:767–773
28. Christensen B, Nielsen J (1999) Metab Eng 1:282–290
29. Pedersen H, Christensen B, Hjort C, Nielsen J (2000) Metab Eng 2:34–41
30. Bertalanffy L (1950) Br J Phil Sci 1:139–164

Received: January 2001

Metabolic Engineering for L-Lysine Production by *Corynebacterium glutamicum*

A. A. de Graaf, L. Eggeling, H. Sahm

Institut für Biotechnologie 1, Forschungszentrum Jülich, 52425 Jülich, Germany
e-mail: a.de.graaf@fz-juelich.de; e-mail: l.eggeling@fz-juelich.de; e-mail: h.sahm@fz-juelich.de

Corynebacterium glutamicum has been used since several decades for the large-scale production of amino acids, esp. L-glutamate and L-lysine. After initial successes of random mutagenesis and screeening approaches, further strain improvements now require a much more rational design, i.e. metabolic engineering. Not only recombinant DNA technology but also mathematical modelling of metabolism as well as metabolic flux analysis represent important metabolic engineering tools. This review covers as state-of-the-art examples of these techniques the genetic engineering of the L-lysine biosynthetic pathway resulting in a vectorless strain with significantly increased dihydrodipicolinate synthase activity, and the detailed metabolic flux analysis by ^{13}C isotopomer labelling strategies of the anaplerotic enzyme activities in *C. glutamicum* resulting in the identification of gluconeogenic phospho*enol*pyruvate carboxykinase as a limiting enzyme.

Keywords. Metabolic engineering, *Corynebacterium glutamicum*, Chromosomal genetic engineering, Metabolic flux analysis, Isotopomer analysis

1 **Introduction** 10

2 ***Corynebacterium glutamicum* and Amino Acids** 11

3 **Anaplerotic Reactions** 11

3.1 Structure of the Anaplerotic Network 11
3.2 Net C3-Carboxylating Flux in vivo 12
3.3 Flux Analysis by ^{13}C Labelling and NMR 13
3.4 C3-Carboxylating and C4-Decarboxylating Flux in vivo 14
3.5 Detailed Flux Information from ^{13}C Isotopomer Analysis 15
3.6 All Anaplerotic Fluxes Resolved in vivo 17
3.7 Anaplerotic Cycling in *Corynebacterium glutamicum* 21

4 **L-Lysine Synthesis** 22

4.1 Control by Aspartate Kinase and Lysine Exporter 22
4.2 Control by Dihydrodipicolinate Synthase Activity 23
4.3 Flux Increase by Engineering *dapA* Expression 25

5 **Conclusion** 27

References 27

Advances in Biochemical Engineering/
Biotechnology, Vol. 73
Managing Editor: Th. Scheper

Abbreviations

HSQC	Heteronuclear Single Quantum Coherence
NMR	Nuclear Magnetic Resonance
OAADc	Oxaloacetate decarboxylase
PEP	Phospho*enol*pyruvate
PEPCk	Phospho*enol*pyruvate carboxykinase
PEPCx	Phospho*enol*pyruvate carboxylase
PyrCx	Pyruvate carboxylase

1 Introduction

The bacterium *Corynebacterium glutamicum* is an example of a microorganism of which the cellular metabolism is engineered for more then 40 years [1] now. After its discovery as an L-glutamate-excreting bacterium, mutant strains useful for the fermentative production of L-glutamate on a large scale were breeded [2]. The successfull development of such producer strains was largely an iterative procedure involving basically two steps. Mutagenesis and small scale fermentations were performed to choose among hundreds of strains the individual strain with the highest productivity. This latter strain was then again subjected to mutagenesis followed by small scale fermentations to choose again the best individual strain. These steps were repeated several times to generate a line of strains, actually a dynasty, with increased flux towards L-glutamate. Obviously, very effective producer strains are made by this procedure. The concentrations obtained for L-lysine and L-glutamate exceed 150 g l^{-1} with yields over 0.5 g g^{-1} [2, 3]. Basically the same procedure was applied to obtain mutants producing other amino acids, too [4].

Whereas this procedure of strain development was more or less dependent on chance, strain development is now shifting to a more rational design, termed metabolic engineering. The recent construction of an L-isoleucine-producing strain from *C. glutamicum* [5] represents a good example of a rational strain design based on a more or less classical engineering approach [6, 7]. In cases where a straightforward approach is less obvious, metabolic engineering strategies must be directed towards the entity of the cell with all its fluxes, reactions and structures. Obviously, this requires the true merging of a whole set of different biochemical, genetical, physical, and mathematical techniques. It serves to (i) increase knowledge of the relevant steps and mechanisms of product fluxes, (ii) to combine this knowledge with classically obtained strains for their further development, and (iii) to rapidly develop new producer strains. L-lysine synthesis with *C. glutamicum* represents a highly illustrative example where knowledge has very much advanced in recent years due to the integrated application of techniques from different fields including biochemistry [8, 9], genetics [10, 11] as well as mathematical modelling and flux analysis [12]. Therefore, in the present review we will take L-lysine synthesis in *C. glutamicum* as an example to illustrate the most recent developments of quantifying fluxes in the central metabolism by sophisticated NMR-approaches, as well as the molecular engineering of the chromosome.

2
Corynebacterium glutamicum and Amino Acids

The use of proteinogenic amino acids and their estimated quantities produced are given in [4]. The largest volumes made are that of L-glutamate, L-lysine and D,L-methionine, with currently 900,000, 420,000, and 350,000 tonnes per year, respectively. L-glutamate and L-lysine are exyclusively made by mutants of *C. glutamicum.* This organism is a Gram-positive non-sporulating bacterium which can be isolated from soil. Very closely related bacteria are *C. melassecola, Brevibacterium thiogenitalis, B. lactofermentum* and *B. flavum*, the latter two organisms being proven to be subspecies of *C. glutamicum* [13]. These bacteria belong together with *Mycobacterium* and *Nocardia* species to the CMN subgroup of Gram-positive bacteria, which is characterized by a special outer lipid layer within the cell envelope containing mycolic acids which are branched fatty acids and which are thought to contribute significantly to the permeability of the cell wall [14]. The genome size of *C. glutamicum* is 3309 kb [15], and the entire sequence has been established. This, as well as the whole set of sophisticated methodology to enable directed in vivo mutagenesis, like gene exchange [16–18], or transposon mutagenesis [11, 19] makes the organism an ideal object of rapid and directed molecular engineering to deepen knowledge and improve metabolite production. *Corynebacterium glutamicum* also has been the subject of a number of studies which are in the forefront of the development of metabolic flux analysis techniques and applications of metabolic engineering [12, 20, 21]. This will be illustrated in this contribution by recent results obtained in the study of the anaplerotic reactions as well as the L-lysine biosynthetic pathway in *C. glutamicum.*

3
Anaplerotic Reactions

3.1
Structure of the Anaplerotic Network

A particular fascinating target of metabolic engineering of *Corynebacterium glutamicum* is the set of anaplerotic reactions. In the anaplerotic node the two precursor metabolites oxaloacetate and pyruvate are generated, which form the basis for as much as 35% of the cell material, not considering the obvious relevance of pyruvate-generated acetyl-CoA to generate ATP via oxidation in the citric acid cycle. These two metabolites form the backbone of L-glutamate and L-lysine. Despite the obvious relevance on the proper supply of oxaloacetate and pyruvate in high-level producer strains, knowledge of the fluxes in the anaplerotic knode was surprisingly limited. Only a gene-directed inactivation of the phospho*enol*pyruvate carboxylase revealed that this enzyme activity is neither essential for growth nor for amino acid production [22]. A subsequent

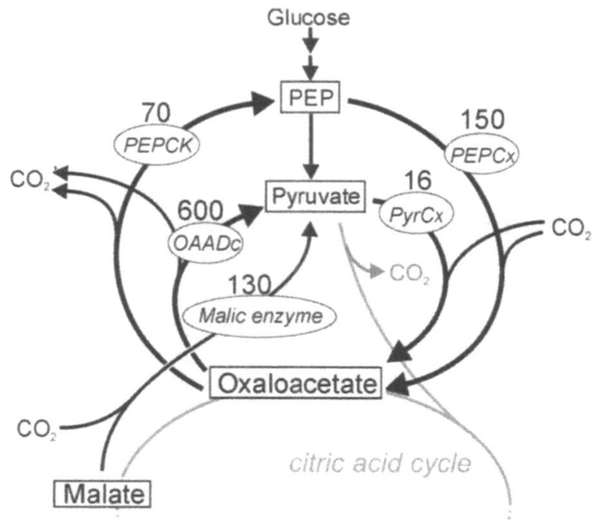

Fig. 1. The diversity of anaplerotic enzymes present in *Corynebacterium glutamicum*. Numbers next to enzyme names represent typical in vitro activities in mU mg protein^{-1}. Abbreviations: PEPCk, phospho*enol*pyruvate carboxykinase; OAADc, oxaloacetate decarboxylase; PyrCx, pyruvate carboxylase; PEPCx, phospho*enol*pyruvate carboxylase

^{13}C-NMR study, which enabled $^{13}CO_2$ incorporation to be traced, gave definite proof of the presence of a second carboxylating reaction in *C. glutamicum* [23]. The investigation of this enzyme activity resulted in the detection of pyruvate carboxylase activity [24] and the cloning of its gene [25]. Thus, *C. glutamicum* has the pyruvate dehydrogenase shuffling acetyl-CoA into the citric acid cycle, and pyruvate carboxylase together with phospho*enol*pyruvate carboxylase supplying oxaloacetate for anaplerotic purposes, as well as the decarboxylating enzymes oxaloacetate decarboxylase, phospho*enol*pyruvate carboxykinase and malic enzyme ([26, 27] (Fig. 1). This is a surprising diversity of enzymes in comparison to other organisms, since *E. coli*, for instance, has as carboxylating enzyme only the phospho*enol*pyruvate carboxylase, and *Bacillus subtilis* only the pyruvate carboxylase. The question for the true *in vivo* activities of all these enzymes in *C. glutamicum* naturally arises. Obviously, neither specific activities determined via *in vitro* enzyme tests (Fig. 1) nor results of genetic studies can be used to identify the actual fluxes in this complex subset of metabolism [6]. Instead, refined methods for Metabolic Flux Analysis have to be applied.

3.2
Net C3-Carboxylating Flux in vivo

To determine the in vivo fluxes, Metabolic Flux Analysis techniques combining metabolite balancing and stable isotope labelling must be applied. Metabolite balancing provides the first step in flux quantitation by using mass balances of intracellular metabolites at steady state [28, 29]. Thus, for the anaplerotic node

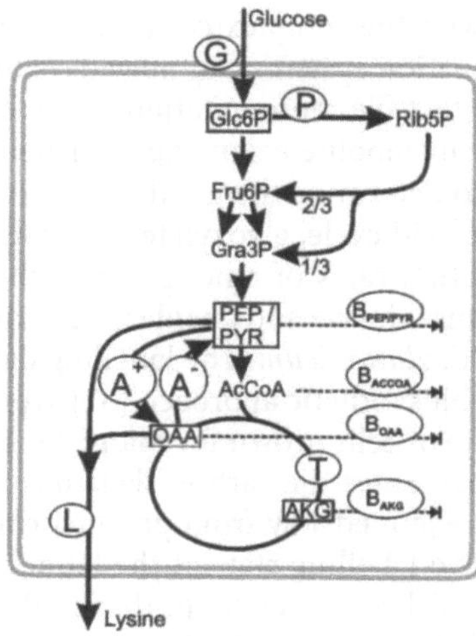

Fig. 2. Principle of net anaplerotic (i.e. C3-carboxylating (A^+) minus C4-decarboxylating (A^-)) flux determination in L-Lysine producing *C. glutamicum* via metabolite balancing. The glucose uptake rate G and the lysine production rate L are measured, and the main precursor drain-offs for biopolymer synthesis $B_{PEP/PYR}$, B_{ACCOA}, B_{OAA} and B_{AKG} are caculated from the biomass composition and the growth rate. The metabolite balance for the oxaloacetate pool can be written as: $T\text{-}B_{AKG} + A^+ = T + A^- + L + B_{OAA}$, i.e. the net anaplerotic flux $A^+ - A^- = L + B_{AKG} + B_{OAA}$. Abbreviations: T, activity of the citric acid cicle; P, activity of the oxidative pentose phosphate pathway; Glc6P, glucose 6-phosphate; Fru6P, fructose-6-phosphate; Gra3P, glyceraldehyde 3-phosphate; PEP, phosphoenolpyruvate; PYR, pyruvate; AcCoA, acetyl coenzymeA; OAA, oxaloacetate, AKG, 2-oxoglutarate

in *Corynebacterium glutam*icum the metabolite balancing approach allows to determine the net C3 carboxylating (i.e. total C3-carboxylating minus C4-decarboxylating) activity once the glycolytic flux at the level of phosphoglucose isomerase or enolase is known (Fig. 2). Resolution of the forward and reverse fluxes by metabolite balancing is not possible, unless other constraints such as energy considerations or cofactor balances are incorporated in the calculation [28, 29]. However, application of this type of constraints may be questionable since first, P/O stoichiometries may vary depending on physiological conditions and secondly, unknown processes may influence cofactor balances [12, 30].

3.3
Flux analysis by ^{13}C labelling and NMR

The second step, i.e. resolution of the forward and reverse fluxes, can be accomplished by stable isotope labelling procedures. Many applications of ^{13}C labelling have been described [12, 31, 32], but also ^{15}N labelling [33] has been used to study metabolic fluxes in *C. glutamicum*. In the case of ^{13}C labelling, a closed balance for each carbon of each intracellular metabolite representing a node in the metabolic network is formulated, whereby the metabolite balancing is automatically integrated in the approach. This model together with the measured positional ^{13}C enrichments is then used to derive the fluxes [12, 34]. In order to efficiently deal with the complexity of truly comprehensive metabolic models, the necessary equation systems can nowadays be automatically generated by computer from a text file representation of the metabolic network [35, 36]. Using this integrated metabolite balancing/^{13}C labelling approach, Marx et al. [12, 31, 32] were able to determine the net fluxes through glycolysis, pentose phosphate pathway, citric acid cycle, glyoxylate pathway, lysine biosynthesis as well as bidirectional reaction rates of among others the phosphoglucose isomerase, transketolase, transaldolase and anaplerotic carboxylation reactions in vivo in several strains of *C. glutamicum*. The labelling data in these studies was derived, following a retrobiosynthetic approach [37], from amino acids isolated from a protein hydrolysate of cells grown for many doubling times in the presence of ^{13}C-labeled glucose. Since the carbon skeletons of amino acids are derived in a well-known, predefined way from precursor metabolites of the central metabolism, the in vivo labelling state of the latter can be concluded from that of the amino acids, which act as a storage device. Rather than by ^{13}C NMR, the positional ^{13}C enrichments of metabolic intermediates are most conveniently analysed by proton NMR due to the fact that ^{13}C-bonded protons produce signals that are easily distinguishable from ^{12}C-bonded protons (Fig. 3a). A convenient procedure is to record two spectra for each sample, one without, the other with broadband ^{13}C decoupling. The difference spectrum enables the distortionless quantitation of the ^{13}C-coupled proton signals [38]. This approach requires individual metabolites to be purified from the fermentation broth or protein hydrolysate since otherwise heavy peak overlap will prevent analysis.

3.4
C3-Carboxylating and C4-Decarboxylating Flux in vivo

The basis for the simultaneous quantitation of the C3-carboxylating and C4-decarboxylating fluxes in microorganisms by ^{13}C labelling is elucidated in Fig. 4. Thus, in experiments employing [1-^{13}C]glucose or [6-^{13}C]glucose an elevated positional labelling of pyruvate C-2 will be observed upon the presence of C4-decarboxylating activity. The ^{13}C enrichments of pyruvate C-2 and C-3 as well as oxaloacetate C-2 and C-3, together with the metabolite balances, can be used to calculate forward and reverse anaplerotic fluxes (Fig. 4). Representative literature data for *C. glutamicum* compiled in Table 1 show that using these techniques, excessive substrate cycling has consistently been observed in this organism under a variety of conditions. This suggests that the C4-decarboxylating activity is constitutively expressed in *C. glutamicum* even during growth on glucose. Thus, while substrate cycling in *E. coli* was recently demonstrated to occur only in glucose-limited chemostat culture [39], the situation in *C. glutamicum* is clearly different.

Data obtained from flux analyses of isogenic strains of *C. glutamicum* in chemostat cultures revealed a remarkably strong correlation between L-lysine production and C4-decarboxylating activity as well as C3-carboxylating activity (Fig. 5). These results suggested that elimination of the C4-decarboxylating activity and/or overexpression of C3-carboxylating activity via recombinant

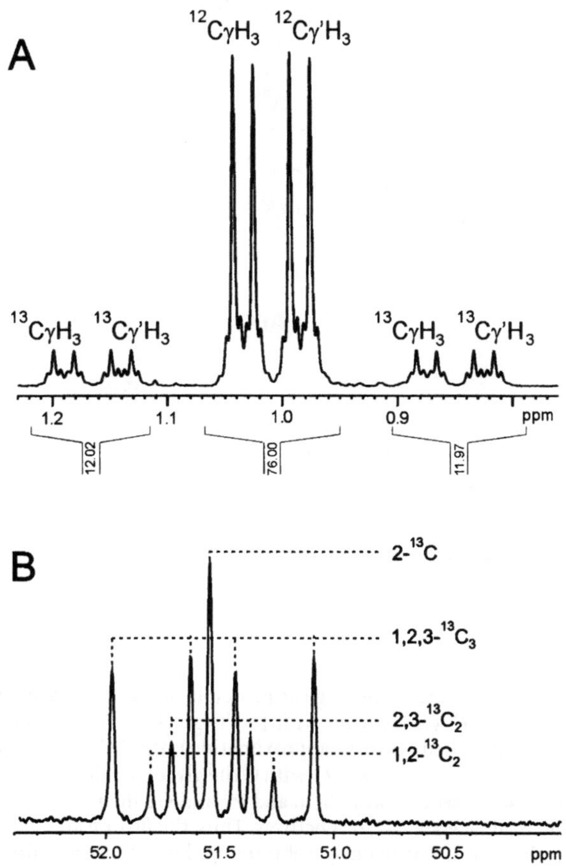

Fig. 3. a Determination of positional ^{13}C enrichment (24%) in the methyl groups of ^{13}C-labelled valine purified fron a biomass hydrolysate of *C. glutamicum* by integration of the ^{13}C satellites in the ^{1}H NMR spectrum; (b) measurement of relative isotopomer abundancies from singlet, doublet (2 varieties) and doublet-of-doublets signals in the ^{13}C NMR spectrum of ^{13}C-labelled alanine

DNA technology might improve lysine yield in *C. glutamicum*. Therefore, it was of prime interest to quantitate the in vivo flux through each of the enzyme reactions potentially involved in the C3-C4 interconversion in *C. glutamicum* in order to decide which enzyme is responsible. For this purpose, a completely new labelling strategy based on ^{13}C isotopomer analysis was developed.

3.5 Detailed Flux Information from ^{13}C Isotopomer Analysis

When using uniformly labelled substrates such as [$^{13}C_6$]glucose against a background of unlabelled substrate, positional ^{13}C enrichments do not contain any information on the fluxes [40, 41]. Instead, this information is contained in the relative abundancies of differently sized fragments of the original glucose ^{13}C

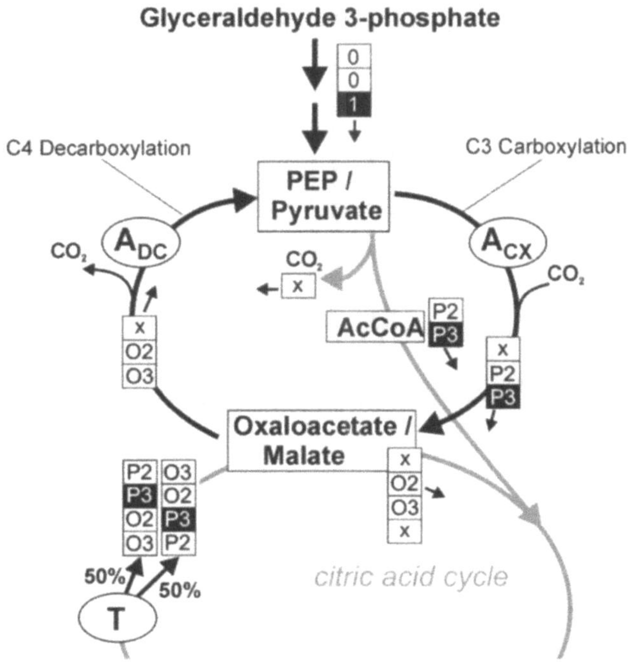

Fig. 4. Carbon-13 labelling routes revealing the principle of C4-C3 backflux identification by NMR upon incubation of *C. glutamicum* with [1-^{13}C]glucose. Considering that C-2 of the triose phosphates is unlabelled (to first approximation), the ^{13}C balance for pyruvate C-2 reads $P2 \times (G + A_{DC}) = (G \times 0 + A_{DC} \times O2)$, with G the glycolytic flux at the level of the triose phosphates, A_{DC} the C4-decarboxylating flux, and P2 and O2 the positional ^{13}C enrichments of pyruvate C-2 and oxaloacetate C-2, respectively. Thus, the flux ratio $A_{DC}/(G + A_{DC})$ is equal to the measured ratio P2/O2. Analogously, subtracting the ^{13}C balances for oxaloacetate C-2 and C-3 yield $(O3\text{-}O2) \times (T + A_{CX}) = (P3\text{-}P2) \times A_{CX}$, with T the activity of the citric acid cycle, A_{CX} the C3-carboxylating flux, and P3 and O3 the positional ^{13}C enrichments of pyruvate C-3 and oxaloacetate C-3, respectively. Thus, the flux ratio $A_{CX}/(T + A_{CX})$ is equal to the measured ratio (O3-O2)/(P3-P2)

Table 1. Substrate cycling in the anaplerotic reactions of *C. glutamicum* (expressed as % of the molar glucose uptake rate) undar various cultivation conditions as identified by ^{13}C labelling-based flux analysis

Strain	Cultivation	Product	Net anaplerosis	C3-carboxylating flux	C4-decarboxylating flux	Ref.
ATCC 13032	Batch	–	23	72	49	[69]
MH20–22B	Chemostat	lysine	38	69	31	[12]
LE4	Chemostat	–	24	96	72	[31]
LE4	Chemostat	glutamate	29	47	18	[31]

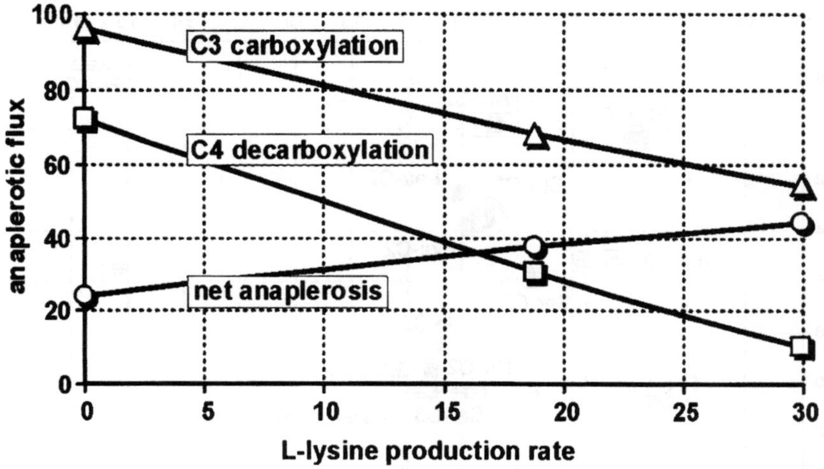

Fig. 5. Flux analysis data [12, 31, 32] on C3-C4 conversions in isogenic strains derived from lysine-producing *C. glutamicum* MH20–22B in chemostat cultures revealing a strong correlation with the L-lysine production rate. Rates are molar and expressed as % of the glucose uptake rate

backbone in the products of metabolism. These fragments can be elegantly detected by direct ^{13}C NMR due to the fact that neighboring ^{13}C nuclei produce multiplet hyperfine splittings of the resonance lines in the NMR spectrum (Fig. 3b). Although the chemical shift dispersion of ^{13}C is much larger than that of ^{1}H, this procedure also requires the metabolites to be at least partially purified from a protein hydrolysate. This drawback has been overcome by newest 2-dimensional Heteronuclear Single Quantum Coherence (HSQC) NMR experiments which allow to analyse the isotopomer distributions of all amino acids in a protein hydrolysate in a single experiment [42]. In the resulting 2D spectrum, the ^{13}C multiplet hyperfine structures are dispersed according to the chemical shift of the proton directly bonded to the carbon (Fig. 6). Thus, this type of NMR experiment has an extremely high information content. Moreover, it considerably simplifies the experimental work by obviating the need to purify the single amino acids from the hydrolysate as required [12] for positional enrichment studies. Recently, it was shown that the analysis by GC-MS of a protein hydrolysate yields a comparable, yet complementary, information content while offering much better sensitivity than NMR [43]. Highly efficient and versatile mathematical modelling procedures [40, 41, 44] allow to extract the flux information also from the complex isotopomer data set.

3.6
All Anaplerotic Fluxes Resolved in vivo

The earlier flux analyses based on positional ^{13}C enrichment patterns did not succeed in resolving the two C3-carboxylating enzymes phospho*enol*pyruvate carboxylase (PEPCx) and pyruvate carboxylase (PyrCx) because the carbon routes in both reactions to oxaloacetate are identical and because no differences

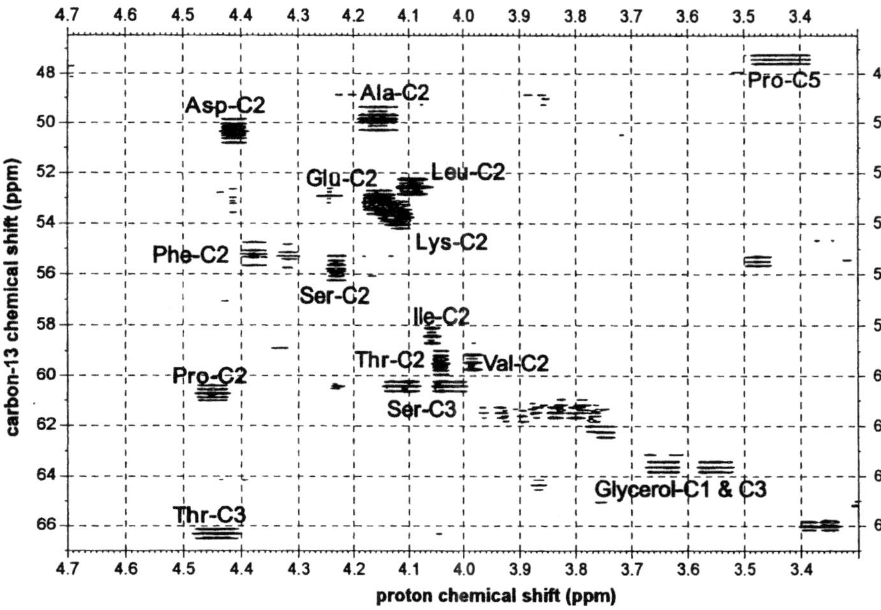

Fig. 6. Detail of a 2D HSQC contour lines spectrum of a protein hydrolysate of *Corynebacterium glutamicum* showing the C_α-resonances of several amino acids as indicated. Cross-sections along the ^{13}C chemical shift dimension yield multiplets as in Fig. 3b that can be used to determine relative isotopomer abundancies. The cells were incubated with $[^{13}C_6]$glucose against a background of both unlabelled and $[1\text{-}^{13}C]$glucose

in labelling of PEP and pyruvate could be observed. Therefore, in a recent analysis based on isotopomer labelling patterns we employed an optimised mixture of labelled substrates (Fig. 7) in which a co-feeding of $[3\text{-}^{13}C]$lactate was applied in order to induce a differential labelling of the PEP and pyruvate pools. The substrate mixture further contained glucose of which 10% was $[^{13}C_6]$glucose, commonly used in isotopomer analysis [42, 45], applied against a background of 90% primarily unlabelled glucose. Since lactate represented only 8.5% of the total carbon source and the experiments were conducted under C-limiting conditions, i. e. no measurable levels of glucose and lactate were observed, it can be expected that the metabolism was effectively undisturbed as compared to the situation of glucose being the sole carbon source. As can be seen in Fig. 8a and b, the influx of $[3\text{-}^{13}C]$lactate was indeed found to lead to pyruvate with a significantly higher abundance of isotopomers labelled in C-3 but not in C-2 as compared to PEP.

Since *Corynebacterium glutamicum* does not possess a PEPsynthetase, no $[3\text{-}^{13}C]$PEP isotopomers can be formed from pyruvate. Thus, any $[3\text{-}^{13}C]$oxaloacetate isotopomers must result from the action of PyrCx in vivo and their relative abundance allows to quantitate the relative contributions of PEPCx and PyrCx to oxaloacetate synthesis. In the aspartate derived from oxaloacetate a content of isotopomers labelled in C-3 but not in C-2 similarly high as that in pyruvate was found (Fig. 8c), suggesting synthesis of oxaloacetate from pyru-

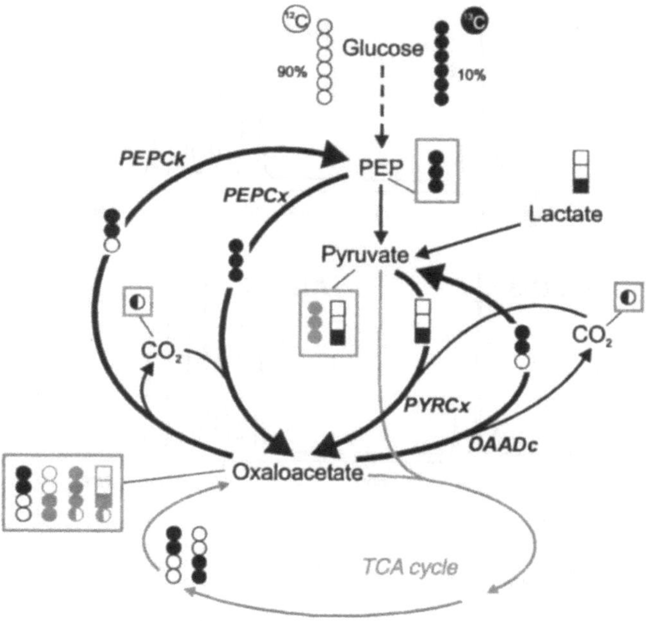

Fig. 7. Rationale for the detailed quantification of C3-C4 interconversion in *C. glutamicum* by isotopomer analysis explained in the text. The substrate mixture consists of unlabelled and uniformly labelled glucose, and [3-^{13}C]lactate as a co-substrate. Full circles/boxes represent ^{13}C carbons, empty ones ^{12}C. Isotopomers crucial for enzyme activity identification are in black, abundant secondary ones shaded grey

vate rather than from PEP. This qualitative view was completely confirmed by the ensuing precise mathematical analysis, which showed that 89% of anaplerotic oxaloacetate synthesis is via PyrCx, and only 11% via PEPCx [46]. Thus, in carbon-limited, glucose-grown chemostat cultures of *C. glutamicum* pyruvate carboxylase is the principal anaplerotic reaction. This contrasts with earlier assumtions that PEPCx was the principal route, but confirms another study that investigated relative use of PEPCx and PyrCx using ^{13}C NMR and GC-MS [47].

While the question of relative use of PEPCx and PyrCx could have been solved from analysis of positional ^{13}C enrichments alone upon co-feeding of [3-^{13}C]lactate, isotopomer analysis involving more complex measurement and modelling procedures was necessary to differentiate between the various C4-decarboxylating enzyme activities present in *C. glutamicum*. Therefore, the second purpose of the new labelling strategy was to produce a unique isotopomer composition of TCA-cycle-generated oxaloacetate in order to detect its back-cycling to PEP and/or pyruvate. The [$^{13}C_6$] glucose applied against a background of unlabeled glucose, if metabolised exclusively via glycolysis, gives rise only to [$^{12}C_3$] and [$^{13}C_3$] isotopomers in PEP and pyruvate. If glucose 6-phosphate is metabolised over the oxidative pentose phosphate pathway and the transaldolase/transketolase routes, it can be shown that not only the [$^{12}C_3$] and

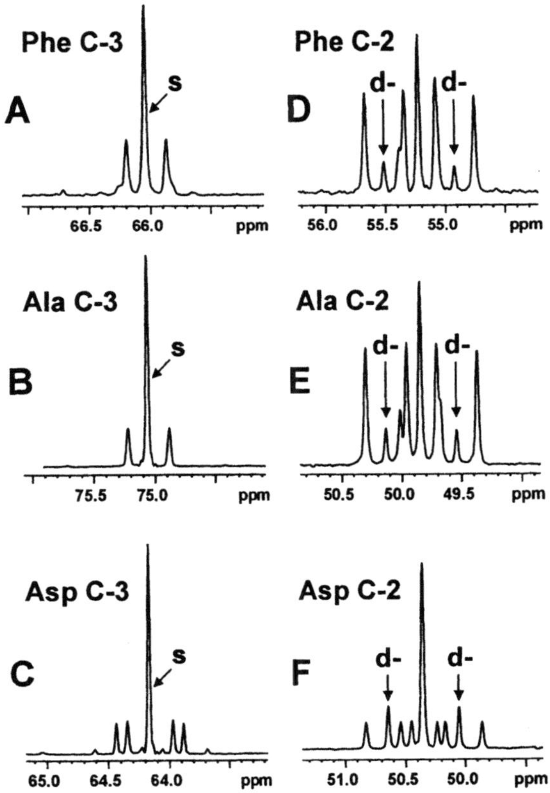

Fig. 8. Experimental ^{13}C NMR spectra of several amino acids from a hydrolysate of *C. glutamicum* ATCC 13032 incubated with a mixture of ^{13}C-labelled glucose and lactate (see Fig. 7) (a) the C-3 carbon of phenylalanine (Phe), (b) the C-3 carbon of alanine (Ala), (c) the C-3 carbon of aspartate (Asp), (d) the C-2 carbon of Phe, (e) the C-2 carbon of Ala, (f) the C-2 carbon of Asp. Cf. Fig. 3b. The elevated singlet (s) contribution in Ala C-3 as compared to Phe C-3 reflects the influx of [3-^{13}C]lactate into the pyruvate pool; the similarly high singlet contribution to Asp C-3 indicates that the main anaplerotic activity is by pyruvate carboxylase and not PEPcarboxylase. The Asp C-2 signals reveal the high abundance of the [1,2-$^{13}C_2$] isotopomer, identified from the d- doublet signals as indicated, resulting from citric acid cycle activity (Fig. 7). The significant presence of this isotopomer in Phe seen from the C-2 d-doublet signals is evidence of a strong oxaloacetate-decarboxylating flux via PEPcarboxykinase. Since the [1,2-$^{13}C_2$] abundance in Ala is virtually identical to that in Phe, it is concluded that litle or no activity of oxaloacetate decarboxylase and malic enzyme is present in vivo

[$^{13}C_3$] isotopomers, but also the [2,3-$^{13}C_2$] and the [1-^{13}C] isotopomers of PEP and pyruvate will be generated. The important point is, however, that no [1,2-$^{13}C_2$] PEP or pyruvate will be formed via any of these routes. In contrast, metabolisation via pyruvate dehydrogenase and the citric acid cycle of the most prominent [$^{13}C_3$]pyruvate isotopomer against the background of unlabeled pyruvate leads primarily to [1,2-$^{13}C_2$] and [3,4-$^{13}C_2$] isotopomers of oxaloacetate (Fig. 8f). After decarboxylation by PEPcarboxykinase (PEPCk) or oxaloacetate decarboxylase (OAADc) these give rise to [1,2-$^{13}C_2$]PEP and [3-^{13}C]PEP or

[1,2-$^{13}C_2$] pyruvate and [3-^{13}C] pyruvate (Fig. 7). Considerable amounts of the [1,2-$^{13}C_2$] isotopomers uniquely reflecting oxaloacetate decarboxylation were found in the experimental spectrum of the C-2 carbon of phenylalanine (Fig. 8 d), indicating a strong backflux of oxaloacetate to PEP via the action of PEPCk. The spectrum of carbon 2 of pyruvate-derived alanine (Fig. 8 e) showed an abundance of [1,2-$^{13}C_2$] isotopomers almost identical to that in phenylalanine. Therefore, it was concluded that negligible recycling of oxaloacetate (and/or malate) to pyruvate occurred in *C. glutamicum* under the conditions studied. Thus, despite their high in vitro activities OAADc and malic enzyme appeared to be inactive in vivo. This illustrates the usefulness and added value of the ^{13}C NMR isotopomer analysis.

3.7 Anaplerotic Cycling in *Corynebacterium glutamicum*

The final flux distribution over the anaplerotic enzymes of *C. glutamicum* resulting from the isotopomer analysis [46] is shown in Fig. 9. The activity of PEPCk even during growth at a rate of 0.1 h^{-1} on glucose leads to a futile cycle in which the energy equivalent of approx. 1 mmol ATP per gram dry weight and hour is dissipated. Since the biomass synthesis at this growth rate may require around 3.5 mmol ATP per gram dry weight and hour [48] it is to be expected that this substrate cycling adds significantly to the maintenance energy requirement of *C. glutamicum*. Furthermore, the fact that the PEPCk reaction

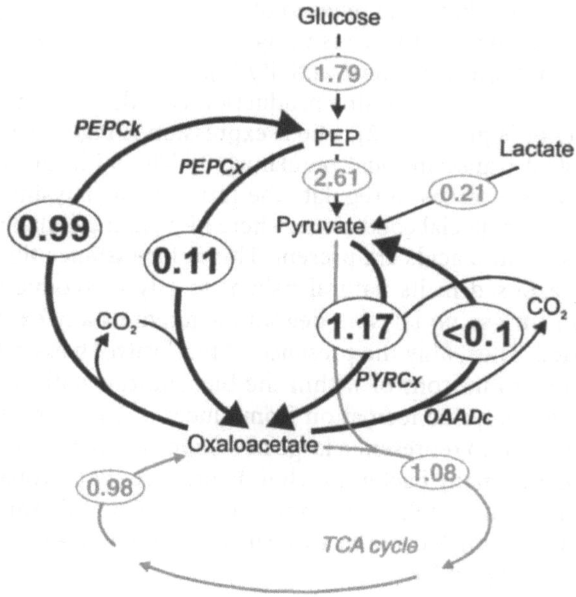

Fig. 9. Finally determined detailed flux distribution in the anaplerosis of *C. glutamicum* ATCC 13032. Data from [46]

withdraws aspartate which is essential for lysine biosynthesis may limit lysine yields. Therefore, construction of a *C. glutamicum* strain with strongly reduced anaplerotic cycling activity by deleting the gene for PEPCk is an important target for metabolic engineering of *C. glutamicum* towards increased growth efficiency and improved lysine yields.

4 L-Lysine synthesis

4.1 Control by Aspartate Kinase and Lysine Exporter

As mentioned, the biosynthesis of lysine from oxaloacetate and pyruvate in *C. glutamicum* occurs via a split pathway [38] (Fig. 10). Clearly, a split pathway is untypical for a biosynthesis pathway. It has been demonstrated that in addition to L-lysine formation this pathway structure ensures the reliable provision of the cell with the intermediate D,L-diaminopimelate, which is an important linking unit within the peptidoglycan layer [49]. One step of flux control through the pathway is at the level of the aspartate kinase. As is typical of an enzyme at the start of a lengthy synthesis pathway, the kinase is controlled in its catalytic activity. The enzyme activity is allosterically inhibited when L-lysine plus L-threonine together are present in excess. Due to its importance in flux control in one line of producers this feed back control is removed by mutations in the β-subunit of the kinase [50]. Also a strain with two copies of the kinase genes was made and shown to result in increased L-lysine accumulation [51]. In another line of stains the kinase is relieved of allosteric inhibition due to low homoserine dehydrogenase activity resulting in a low L-threonine concentration which no longer inhibits the kinase activity [2].

A further flux control for L-lysine production is at the level of export. A specific export carrier is present [52], whose expression is regulated by an autogeneously controlled transcriptional regulator[53]. This hitherto unknown type of control by export, serves to regulate the intracellular L-lysine or L-arginine concentration under special conditions, where high, non cellular-made concentrations of these amino acids are present. This is for instance the case when *C. glutamicum* is exposed in its natural habitat to L-lysine-containing peptides. Since the organism has no L-lysine-degrading activities any excess of L-lysine must be exported. Thus, only the presence of this "valve" has enabled that mutations overcoming flux control within the biosynthesis pathway have indeed resulted in cellular L-lysine formation from glucose. The L-lysine exporter has now been recognized to represent a large new superfamily of translocators with members present in many bacteria including archeae [54]. Probably all are involved in export of small solutes from the cell [55]. One subfamily of the LysE superfamily is RhtB, which contains exporters of *E. coli* related with L-threonine and L-homoserine export [56].

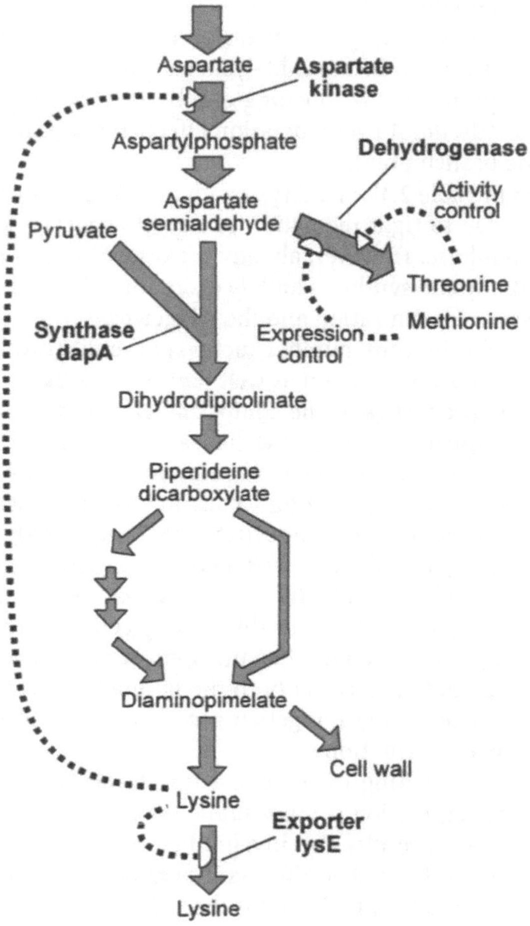

Fig. 10. Schematic representation of the split biosynthetic pathway of L-lysine in wildtype *Corynebacterium glutamicum* including the branch point of aspartate semialdehyde distribution. The metabolites derived from the aldehyde via the synthase activity are D,L-diaminopimelate and L-lysine, whereas that resulting from dehydrogenase activity are L-threonine, L-methionine, and L-isoleucine. The activity of the dehydrogenase is inhibited at elevated L-threonine concentrations and its synthesis is repressed by L-methionine. Accumulating intracellular lysine causes feedback inhibition of aspartate kinase and activates *lysE* transcription

4.2
Control by Dihydrodipicolinate Synthase Activity

A further flux control step within L-lysine synthesis is the aspartate semialdehyde branch point. The aldehyde is either used as a substrate for the homoserine dehydrogenase, or together with pyruvate as a substrate for the dihydrodipicolinate synthase (Fig. 10). Whereas the homoserine dehydrogenase is allosterically controlled in its catalytic activity by the L-threonine concentration and repressed by L-methionine [57], no such control is known for the dihydrodipicolinate synthase [58]. Overexpression of the dihydrodipicolinate synthase gene *dapA* resulted in increased L-lysine accumulation [59]. At first sight this could be interpreted as the "opening of a bottleneck". However, as will be outlined subsequently, *dapA* overexpression effects the flux at the entire aspartate semialdehyde branch point.

As can be seen in Table 2, the wild type with one *dapA*-copy does not excrete L-lysine, which is due to tight regulation of flux at aspartate kinase and dihydrodipicolinate synthase. However, already introduction of a second copy results in increased L-lysine synthesis and its excretion. This is due to an elevated intracellular L-lysine concentration and the triggering of the export machinery. A further increase in the copy number increases the dihydrodipicolinate synthase activity and L-lysine excretion as well [60]. This is due to two effects. The first are the kinetic properties of the competing enzymes at the branch point. Thus the dihydrodipicolinate synthase has a low affinity for the aldehyde ($K_m = 2.08$ mM) and a low maximal specific activity ($v_{max} = 0.09$ µmol min^{-1} (mg protein)$^{-1}$, whereas the corresponding values for the homoserine dehydrogenase are nearly one order of magnitude higher ($K_m = 0.37$ mM; $v_{max} = 0.75$ µmol min^{-1} (mg protein)$^{-1}$. These data, as well as the concentration of the aspartate semialdehyde in the cell of about 0.05 mM, show that the flux towards L-lysine is determined by the low affinity of the dihydrodipicolinate synthase. Since this flux control could not be operative when the homoserine dehydrogenase would have low affinity and activity, in fact both the homoserine dehydrogenase and the dihydrodipicolinate synthase together are elements of flux control for aspartate semialdehyde distribution.

The second effect resulting in increased flux towards L-lysine as a consequence of *dapA* overexpression is more subtle. As can be seen in Table 2, gradual *dapA* overexpression results also in a gradual reduction of the growth rate. Therefore, cell growth is limited. As the quantification of the intracellular amino acid concentrations revealed (Table 2) it is the L-threonine concentration which is reduced upon *dapA* overexpression. This unexpected finding is confirmed by the fact that addition of L-homoserine, for instance, restores growth of a *dapA* overexpressing strain [60]. Why an expected release of feedback inhibition of

Table 2. The overexpression of *dapA* effects the cellular flux towards L-threonine and L-lysine, as well as the growth rate

C. glutamicum Strain	*dapA* copies	Synthase activity (U mg protein^{-1})	Growth rate (h^{-1})	Intra-cellular Threonine (mmol^{-1})	Intra-cellular Valine (mmol^{-1})	Lysine excretion rate (nmol min^{-1} mg dry wt^{-1})
13032	1	0.05	0.43	9	3	0.0
13032::*dapA*	2	0.082	0.37	3	6	0.2
13032 pKW3::*dapA*	6	0.25	0.36	≈ 1	8	2.7
13032 pJC24	20	0.63	0.22	≈ 1	10	3.8

the homoserine dehydrogenase by L-threonine (Fig. 10) does not compensate for the limitation is not yet understood, but subject to current analysis. However, most importantly, the growth limitation results in an increased availability of intracellular precursors, as for example pyruvate. This is evident from the increased concentration of L-valine (Table 2), which is synthesized from two pyruvate molecules. An additional advantage of increased L-lysine synthesis due to *dapA* overexpression is the reduced extracellular accumulation of some minor byproducts formed [61]. This is the case when *dapA* is overexpressed in the background of a high-level producer strain, like MH20–22B. In this strain, plasmid-encoded *dapA* overexpression results in an increased L-lysine accumulation from about 230 mM to 280 mM, accompanied by a reduction of L-isoleucine and L-alanine from concentrations of 6 mM to concentrations below 1 mM. It should be mentioned that in many amino acid-fermentation processes growth limitations by limiting medium components (e.g. phosphate) are used to achieve increased product formation [62, 63]. Limiting intracellular fluxes by genetic engineering has the advantage to stabilize fermentations by making them independent of variations of limiting medium components.

4.3 Flux Increase by Engineering *dapA* Expression

Due to the importance of the total dihydrodipicolinate synthase activity in high-level producer strains, *dapA* transcription was investigated in detail. This served to finally adjust optimal expression of this gene. This is a particular interesting example of strain engineering to ultimately result in a plasmid-free, self-cloned strain carrying only *C. glutamicum* sequences. *C. glutamicum* promotors are characterized by a –10 region consisting of the consensus sequence TANAAT which is comparable to that of *E. coli*, *Bacillus*, *Lactococcus* or *Streptococcus* [64]. However, in *C. glutamicum* a –35 region is much less conserved. As a first step to engineer the *dapA* promoter, the specific *dapA* transcript initiation site within the *dapB*-orf2-*dapA*-orf4 operon was identified [65]. According to the structure of the synthase protein [66] it has to be concluded that the transcript initiation site is identical or very close to the translation initiation site. This is one of the several examples in *C. glutamicum* where such a situation exists, and is the case, for instance, with *ilvA*, *lpd*, *thrC*, or the *ilvB*-leader. How translation without classical ribosome binding site occurs is unknown. Probably secondary structures around the transcript initiation site are involved as well as additional protein components, like the orf4-encoded polypeptide of the operon. As a second step a deletion analysis of the *dapA* promotor was made, with *dapA* fused to the chloramphenicol acetyl transferase (CAT) gene serving as a reporter [67]. This enabled to confine the essential elements for transcript initiation to an 80 bp fragment (–86 to –7) carrying an essential stretch of T's at position –57 to –52 and of course the essential –10 region.

The further engineering of the promoter and the generation of the respective strain without any vector sequences is outlined in Fig. 11. Based on known sequences of strong promotors of *C. glutamicum* directed mutations were introduced, which among others resulted in promotor MC 20 with nearly four-fold

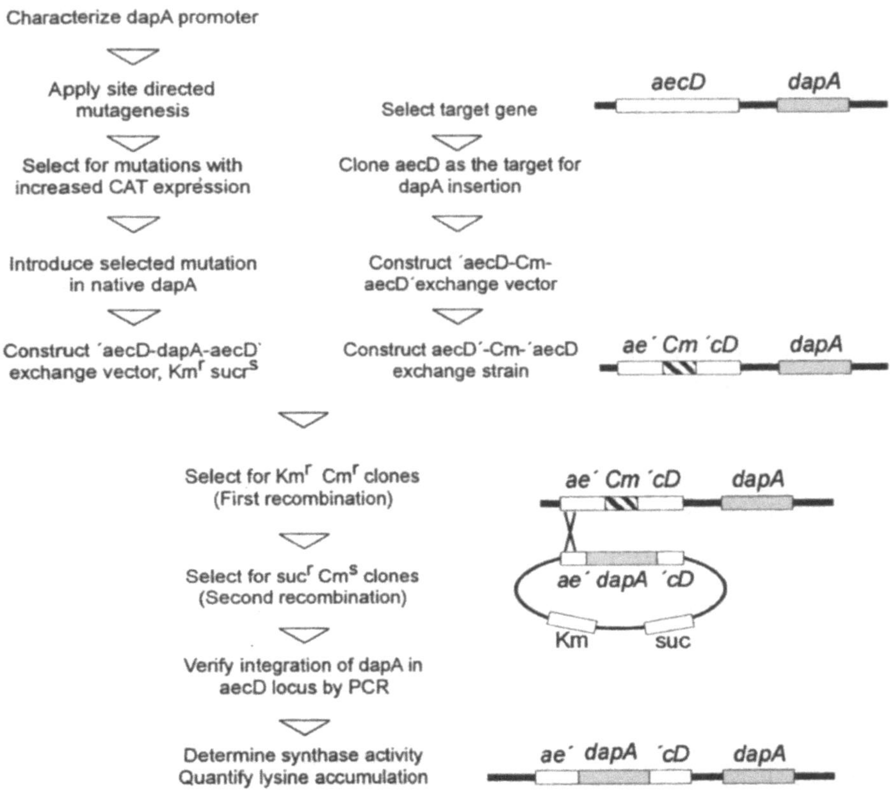

Fig. 11. Flow scheme of the different steps to generate a vectorless L-lysine producer from *C. glutamicum* with increased dihydrodipicolinate synthase activity. This procedure makes extensive use of homologuous recombinations and several positive selection procedures [16]. In the upper left the steps needed to make a *dapA* promoter resulting in high chloramphenicol acetyl transferase activity as well as to generate an exchange vector carrying *dapA* with the mutated promoter and flanking *aecD* regions [68] are given. The middle part of the flow scheme shows the steps to make a strain which has *aecD* interrupted by a Cm^r gene. In the lower part the steps to select for an exchange of the chromosomal Cm^r gene by the mutated *dapA* are illustrated. The proper strain construction is verified by PCR, enzyme measurements and product accumulation. The chromosomal situation of the respective strain is given on the right

Table 3. Effect of *dapA* promotor mutations on dihydrodipicolinate synthase activity and L-lysine accumulation after 48 h of cultivation

Strain	Promotor sequence -31..........................-9	*dapA* copies	Synthase activity (U/mg)	L-Lysine (g/L)
MH20-22B		1	0.046	13.46
MH20-22B	CAAATG......AGGTAACCT	2	0.105	15.98
Mutant MA 16	CAAAATG......AGGTATAAT	2	0.137	16.62
Mutant MC 20	CAAATG......TGGTAACCT	2	0.185	17.27

increased transcription initiation. In the specific MC 20 promotor, A in position -17 is replaced by T (Table 3). Then several steps were necessary to transfer a second copy of *dapA* with the point mutation in its promotor in the chromosome of the L-lysine producer. As the final result, L-lysine producers were obtained with *C. glutamicum* sequences only, which exhibit increased synthase activity. As can be seen from Table 3, the selected strains have a substantially increased L-lysine accumulation.

5
Conclusion

The purposeful metabolic engineering of *Corynebacterium glutamicum* for improved amino acid production was shown to benefit from the integrated application of methods for biochemical analysis, genetic engineering, mathematical modelling and metabolic flux analysis. Chromosomal genetic engineering of the *dapA* resulted in a stable overexpression of dihydrodipicolinate synthase and a significantly increased lysine production as a consequence of redistribution of the fluxes at the aspartate semialdehyde branchpoint in the L-lysine biosynthetic pathway. Detailed analysis of the anaplerotic enzyme activities in vivo by refined ^{13}C isotopomer labelling techniques resulted in the identification of phospho*enol*pyruvate carboxykinase as the enzyme responsible for strong futile cycling in the anaplerotic network of *C. glutamicum* which was previously shown to correlate with a decreased lysine production. In the near future, it is expected that integrated monitoring and modelling of the effects of genetic changes on the proteome, the metabolome and the fluxome level will provide a significantly improved insight in the regulatory processes involved in amino acid overproduction by *Corynebacterium glutamicum.*

Acknowledgement. We thank Degussa and the BMBF for continuous support of the work on *C. glutamicum.*

References

1. Kinoshita S, Udaka S, Shimono M (1957) J Gen Appl Microbiol 3:193
2. Aida K, Chibata I, Nakayama K, Yamada H (1986) Biotechnology of amino acid production. In: Progress in biotechnology, vol 21. Elsevier
3. Eggeling L, Sahm H (1999) Appl Microbiol Biotechnol 52:146
4. Leuchtenberger W (1996) In: Rehm HJ, Reed G (ed) Biotechnology, vol 6, VCH Verlagsgesellschaft, Weinheim, p 455
5. Eggeling L, Morbach S, Sahm H (1997) J Biotechnol 56:167
6. Eggeling L, de Graaf A, Sahm H (1996) Adv Biochem Eng 54:1
7. Eggeling L, Sahm H (1999) In: Lee SY, Papoutsakis ET (ed) Metabolic Engineering, Marcel Dekker, Inc, New York Basel, p 153
8. Jakoby M, Nolden L, Meier-Wagner J, Krämer R, Burkovski A (2000) Mol Microbiol 37:964
9. Peters-Wendisch P, Kreutzer C, Kalinowski J, Patek M, Sahm H, Eikmanns BJ (1998) Microbiology 144:915
10. Schäfer A, Tauch A, Jäger W, Kalinowski J, Thierbach G, Pühler A (1994) Gene 145:69
11. Ankri S, Serebrijski I, Reyes O, Leblon, G (1996) J Bacteriol 178:4412
12. Marx A, de Graaf AA, Wiechert W, Eggeling L, Sahm H (1996) Biotechnol Bioeng 49:111

13. Liebl W, Ehrmann M, Ludwig W, Schleifer KH (1991) Int J Syst Bacteriol 41:225
14. Nikaido H (1994) Science 264:282
15. Bathe B, Kalinowski J, Pühler A (1996) Mol Gen Genet 252:255
16. Schäfer A, Tauch A, Jäger W, Kalinowski J, Thierbach G, Pühler A (1994) Gene 145:69
17. Wohlleben W, Muth G, Kalinowski J (1994) In: Rehm HJ, Reed G, Pühler A, Stadler P (ed) Biotechnology, vol 2, VCH Verlag Chemie, Weinheim, p 457
18. Malumbres M, Mateos LM, Martin JF (1996) In: Hui YH, Khachatourians GG (ed) Food biotechnology: microorganisms. VCH Verlagsgesellschaft, Weinheim, p 423
19. Vertès AA, Inui M, Kobayashi M, Kurusu Y, Yukawa H (1994) Mol Microbiol 11:739
20. Stephanopoulos G, Vallino JJ (1991) Science 252:1675
21. Papoutsakis ET, Lee SY (eds) (1999) Metabolic Engineering. Marcel Dekker, New York Basel
22. Peters-Wendisch PG, Eikmanns BJ, Thierbach G, Bachmann B, Sahm H (1993) FEMS Microbiol Lett 112:269
23. Peters-Wendisch PG, Wendisch VF, de Graaf AA, Eikmanns BJ, Sahm H (1996) Arch Microbiol 165:387
24. Peters-Wendisch PG, Wendisch VF, Paul S, Eikmanns BJ, Sahm H (1997) Microbiol 143:1095
25. Peters-Wendisch PG, Kreutzer C, Kalinowski J, Paték M, Sahm H, Eikmanns BJ (1998) Microbiol 144:915
26. Jetten MS, Pitoc GA, Follettie MT, Sinskey AJ (1994) Appl Microbiol Biotechnol 41:47
27. Lindley ND, Cocaign-Bousquet M, Guyonvarch A (1996) Appl Environ Microbiol 62:429
28. Vallino JJ, Stephanopoulos G (1993) Biotechnol Bioeng 41:633
29. Vallino JJ, Stephanopoulos G (1990) In: Sikdar SK, Bier M, Todd P (ed) Frontiers in bioprocessing. CRC Press, Boca Raton, p 205
30. Schmidt K, Marx A, de Graaf AA, Wiechert W, Sahm H, Nielsen J, Villadsen J (1998) Biotechnol Bioeng 58:254
31. Marx A, Striegel K, de Graaf AA, Sahm H, Eggeling L (1997) Biotechnol Bioeng 56:168
32. Marx A, Eikmanns BJ, Sahm H, de Graaf AA, Eggeling L (1999) Metab Eng 1:35
33. Tesch M, de Graaf AA, Sahm H (1999) Appl Environ Microbiol 65:1099
34. Wiechert W, de Graaf AA (1996) In: Scheper T (ed) Advances in Biochemical Engineering/ Biotechnology, vol. 54: Metabolic Engineering, Springer, Berlin Heidelberg New York, p 111
35. Wiechert W, de Graaf AA (1997) Biotechnol Bioeng 55:101
36. Wiechert W, Siefke C, de Graaf AA, Marx A (1997) Biotechnol Bioeng 55:118
37. Bacher A, Rieder C, Eichinger D, Arigoni D, Fuchs G, Eisenreich W (1999) FEMS Microbiol Rev 22:567
38. Sonntag K, Schwinde J, de Graaf AA, Marx A, Eikmanns BJ, Wiechert W, Sahm H (1995) Appl Microbiol Biotechnol 44:489
39. Sauer U, Lasko DR, Fiaux J, Hochuli M, Glaser R, Szyperski T, Wüthrich K, Bailey JE (1999) J Bacteriol 181:6679
40. Wiechert W, Möllney M, Isermann N, Wurzel M, de Graaf AA (1999) Biotechnol Bioeng 66:69
41. Möllney M, Wiechert W, Kownatzki D, de Graaf AA (1999) Biotechnol Bioeng 66:86
42. Szyperski T (1995) Eur J Biochem 232:433
43. Christensen B, Nielsen J (1999) Metab Eng 1:282
44. Schmidt K, Carlsen M, Nielsen J, Villadsen J (1997) Biotechnol Bioeng 55:831
45. Schmidt K, Nielsen J, Villadsen J (1999) J Biotechnol 71:175
46. Petersen S, de Graaf AA, Eggeling L, Möllney M, Wiechert W, Sahm H (2000) J Biol Chem 275:35932
47. Park SM, Shaw-Reid C, Sinskey AJ, Stephanopoulos G (1997) Appl Microbiol Biotechnol 47:430
48. Gottschalk G (1986) Bacterial metabolism, 2nd edn, Springer, New York
49. Wehrmann A, Phillip B, Sahm H, Eggeling L (1998) J Bacteriol 180:3159
50. Thierbach G, Kalinowski J, Bachmann B, Pühler A (1990) Appl Microbiol Biotechnol 32:443

51. Jetten MSM, Follettie M, Sinskey A (1995) Appl Microbiol Biotechnol 41:76
52. Vrljic M, Eggeling L, Sahm H (1996) Mol Microbiol 22:815
53. Bellmann A, Vrljic M, Pátek M, Sahm H, Krämer R, Eggeling L (2000) submitted
54. Aleshin VV, Zakataeva NP, Livshits VA (1999) TIBS 24:133–135
55. Vrljic M, Garg J, Bellmann A, Wach S, Freudl R, Malecki MJ, Sahm H, Kozina VJ, Eggeling L, Saier MH Jr (1999) J Mol Microbiol Biotechnol 1:327
56. Zakataeva NP, Aleshin VV, Tokmakova IL, Troshin PV, Livshits VA (1999) FEBS Lett 452:228
57. Follettie MT, Shin HL, Sinskey AJ (1988) Mol Microbiol 2:53
58. Cremer J, Treptow C, Eggeling L, Sahm H (1988) J Gen Microbiol 134:3221
59. Cremer J, Eggeling L, Sahm (1991) Appl Environ Microbiol 57:1746
60. Eggeling L, Oberle S, Sahm H (1998) Appl Microbiol Biotechnol 49X:24
61. Kircher M (1998) Mag Soc of Ind Microbiol 48:4
62. Kiss RD, Stephanopoulos G (1991) Biotechnol Prog 7:501
63. Konstantinov KB, Nishio N, Seki T, Yoshida T (1991) J Ferment Bioeng 71:350
64. Pátek M, Bilic M, Krumbach K, Eikmanns BJ, Sahm H, Eggeling L (1997) Biotechnol Lett 19:1113
65. Pátek M, Eikmanns BJ, Pátek J, Sahm H (1996) Microbiol 142:1297
66. Mirwaldt C, Korndörfer I, Huber R (1995) J Mol Biol 246:227
67. Vasicova P, Pátek M, Nesvera J, Sahm H, Eikmanns B (1999) J Bacteriol 181:6188
68. Rossol I, Pühler A (1992) J Bacteriol 174:2968
69. Wendisch VF, de Graaf AA, Sahm H, Eikmanns BJ (2000) J Bacteriol 182:3088

Received: December 2000

Process Development and Metabolic Engineering for the Overproduction of Natural and Unnatural Polyketides

Robert McDaniel[1], Peter Licari[1], Chaitan Khosla[2]

[1] KOSAN Biosciences, Inc., 3832 Bay Center Place, Hayward CA 94545, USA
[2] Departments of Chemical Engineering, Chemistry, and Biochemistry, Stanford University, Stanford CA 94305-5025, USA, e-mail: ck@chemeng.stanford.edu

Polyketide natural products are a rich source of bioactive substances that have found considerable use in human health and agriculture. Their complex structures require that they be produced via fermentation processes. This review describes the strategies and challenges used to develop practical fermentation strains and processes for polyketide production. Classical strain improvement procedures, process development methods, and metabolic engineering approaches are described. The elucidation of molecular mechanisms that underlie polyketide biosynthesis has played an important role in each of these areas over the past few years.

Keywords. Polyketide, Antibiotics, Biosynthesis, Bioprocess development, Metabolic engineering

1 Introduction . . . 32

2 What are Polyketides? . . . 32

3 Why do Natural and Unnatural Polyketides Need to be Overproduced? . . . 32

4 What Properties of a Polyketide Fermentation can be Improved? . . . 34

5 Classical Strain Improvement for Polyketide Production . . . 34

6 Process Development for Polyketide Production . . . 35

6.1 Clone Selection and Culture Preservation . . . 35
6.2 Media Development . . . 36
6.3 Small-Scale Fermentation . . . 37

7 Metabolic Engineering for the Overproduction of Polyketides . . . 39

7.1 Heterologous Expression of Polyketides . . . 39
7.2 Maximizing Gene Expression . . . 40
7.3 Enhancing Precursor Supplies . . . 42
7.4 Superhosts . . . 46
7.5 Genomics Guided Process and Strain Improvement . . . 47

8 Conclusion . . . 48

References . . . 48

Advances in Biochemical Engineering/
Biotechnology, Vol. 73
Managing Editor: Th. Scheper

1 Introduction

Polyketides, a large family of bioactive natural products, have found considerable use in human health and agriculture. Their structural complexity necessitates that they be produced via fermentation routes. Since producing microorganisms typically synthesize relatively small quantities of material, the cost of these substances is unusually high. Therefore, the development of commercially viable processes for polyketide production presents a technologically exciting challenge for the biochemical engineer. A combination of classical genetics, bioprocess engineering, and metabolic manipulation are proving to be effective tools for enhancing the volumetric productivity of polyketide biosynthetic processes. The state of art in each of these areas is summarized in this review.

2 What are Polyketides?

Polyketides are a large family of natural products built from acyl-CoA monomers. These metabolites include many important pharmaceuticals, veterinary agents, and agrochemicals. The enormous structural diversity and complexity of these biomolecules is impressive. Although the actual biological roles of each of these metabolites in the native producing organisms (primarily actinomycetes) are unclear, an extraordinary variety of pharmacological properties have been associated with naturally occurring polyketides. Widely used polyketides include antibacterials (erythromycin, tetracycline, rifamycin), antifungals (amphotericin), anti-cancer agents (doxorubicin), immunosuppressants (FK506, rapamycin), cholesterol lowering agents (lovastatin, compactin), animal health products (avermectin, tylosin, monensin), and agrochemicals (spinosyn). Some of these compounds are shown in Fig. 1.

Polyketides are biosynthesized by large multi-enzyme systems called polyketide synthases (PKSs). These large synthases, which are modular in architecture and function, catalyze step-wise elongation of a polyketide chain, as well as associated functional group modifications. At each step in the chain elongation process, an acyl group monomer is recruited from the available metabolic pool of acyl-CoA precursors. Typical precursors include metabolites such as acetyl-CoA, propionyl-CoA, and other alkyl-CoAs, which are used as chain initiators, and malonyl-CoA and methylmalonyl-CoA, which are used for the elongation process. The size of the polyketide product is controlled by the number of repeated acyl chain extension steps. Throughout its biosynthesis the growing polyketide chain is covalently tethered to the protein assembly. A detailed review of polyketide biosynthesis is presented elsewhere [1, 2].

3 Why do Natural and Unnatural Polyketides Need to be Overproduced?

Although new bioactive microbial polyketides continue to be discovered at a fast rate [3], their subsequent development faces significant hurdles. Among the

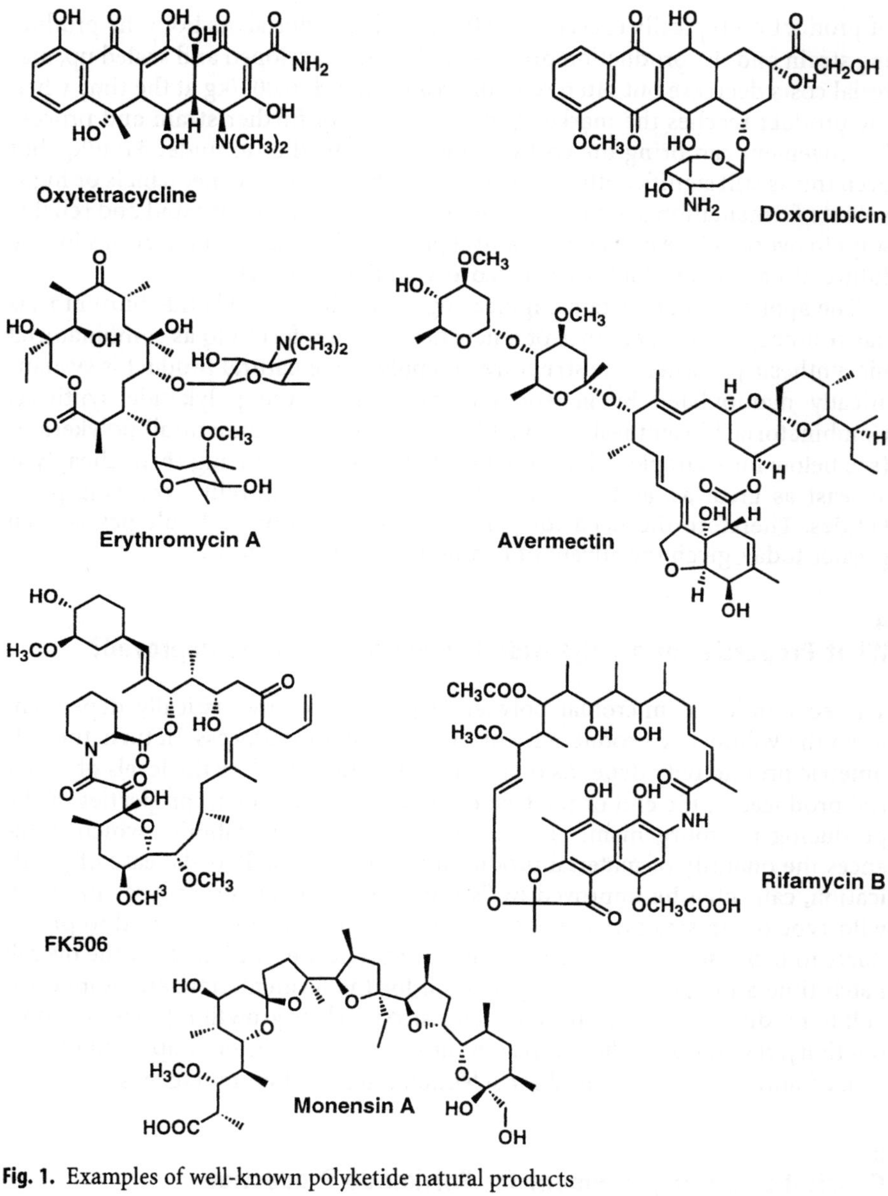

Fig. 1. Examples of well-known polyketide natural products

most significant hurdles is the extraordinarily high cost of producing the bioactive compound, which means that only small amounts of materials are available. This in turn presents a challenge for medicinal chemists, who wish to derivatize the compound to find superior analogs, and for pharmacologists and toxicologists, who seek to study the properties of the natural product (and semi-synthetic derivatives) in assays that require exponentially greater quantities of materials. In the initial stages following the isolation of a novel polyketide, the cost of production typically exceeds $1000/g purified material. As both the producing strain and the production process are further developed and scaled up, material costs decrease but often remain greater than $10,000/kg at the time when the product reaches the market. Several decades of further strain and process improvement can bring the cost of goods down further to under $100/kg, but even this is still significantly more expensive than synthetic medicinals or high-volume fermentation products. Therefore the development of rapid and reliable ways to overproduce polyketide natural products has major implications for the future of natural products drug discovery and development.

The application of protein engineering principles to polyketide biosynthesis has resulted in the emergence of a new field, often referred to as combinatorial biosynthesis [1], where the structure of a polyketide natural product is systematically manipulated by genetic manipulation of the polyketide synthase. Combinatorial biosynthesis has yielded numerous new "unnatural" polyketides (see below for examples); however the challenge of producing them cheaply is at least as great as, and perhaps greater than, for naturally occurring polyketides. Therefore the need for generic overproduction technologies is even greater today, given the advent of combinatorial biosynthesis.

4 What Properties of a Polyketide Fermentation can be Improved?

The economics of microbial polyketide production are critically dependent upon the volumetric productivity of the fermentation process. In turn, the volumetric productivity depends on two factors – the "titer" or the levels of material produced at the end of the fermentation, and the growth properties of the producing microorganism. The titer of a secondary metabolite, which influences the quantity of material produced per batch as well as the ease of purification, can often be improved by "strain improvement" procedures in which wild-type organisms producing secondary metabolites are subjected to procedures to increase titer. Although the growth properties, which affect the overall batch time and operating costs per batch for the fermentation, are more difficult to modify, some desirable characteristics, such as growth rate, specific productivity, resistance to shear forces, reduced aggregation, and improved utilization of nutrients are also used for selection after random mutagenesis.

5 Classical Strain Improvement for Polyketide Production

Currently, the conventional approach for strain improvement involves repeated cycles of random mutagenesis and screening for higher producers. Thus, a wild-type strain is treated with a mutagen, such as nitrosoguanidine, UV, or EMS, and numerous survivors are screened to find several which produce higher amounts of product. The higher producing strains are isolated, again mutagenized, screened, and the process is continued until production is maximized. In practice, the process is usually successful and production levels often increase approximately linearly with time, although it can take many years to achieve multi-gram/liter titers. For example, wild-type *S. erythraea* producing ~100 mg/l of erythromycin has been converted to strains producing 10 g/l (100-fold increase) by repeated cycles of mutagenesis and screening over a period of many years.

A problem in conventional strain improvement is the identification of improved producers during each round of mutagenesis/analysis. The problems reside in the small increase in secondary metabolite observed during each round, the tediousness of (most) assays, and the variability of secondary metabolite production in genetically identical organisms. Typically, thousands of independent clones need to be screened to identify improved producers in each round of mutagenesis. Thus, although strain improvement of producers of polyketides is usually feasible, it requires considerable time and effort, and must be individualized for each strain/product used. Many aspects of conventional strain improvement are often automated today.

6 Process Development for Polyketide Production

An efficient process development program is critical to the success of a metabolic engineering program. In an industrial setting, a process development program often encompasses clone selection, culture preservation, media development, small-scale fermentation development, and ultimately scale-up studies. The objectives of process development programs are to improve volumetric productivity (grams per liter per day), minimize costs, and develop processes that scale. Cost effective and scaleable processes are often realized in processes that are simple and robust. The following provides a brief summary of process development; selected contemporary references are provided as a means to direct the reader to more specific case studies. References provided are not meant to be an exhaustive review of the literature.

6.1 Clone Selection and Culture Preservation

Close interactions between microbiologists, molecular biologists, and process development scientists allow for the rapid transfer of primary colonies or transformants to the process development pipeline. After demonstration that a polyketide of interest is successfully produced from a natural isolate or a recombinant strain, 20–50 colonies are screened under well-defined shake flask conditions. On a practical level, medium and conditions (pH, temperature, and flask configuration) that provide growth and/or production for the host organism are selected in relation to later conditions for large scale production. Even though the nutritional requirements of recombinant derivatives may be different, the selection of these baseline conditions provides an essential starting point.

Although labor intensive, clonal analysis which follows a time-course of growth and production is useful. The frequency at which cultures are sampled is dependent on both the throughput of analysis and the culture volumes, as well as the number of colonies screened. A thorough time-course study is helpful at this stage in that the growth and production kinetics of new clones are unknown. Novel polyketide products have been observed that are unstable in certain fermentation conditions; in these cases, product would not be detected if only end-point screens were employed. Once identified, stability problems may be sometimes avoided by manipulating process conditions. Clone heterogeneity exists both in primary transformants and mutational screens [4]. Screening a large number of clones may provide pertinent information (e.g., plasmid stability in the case of recombinant DNA processes) in addition to productivity. Process development may take advantage of differences that exist in nutritional requirements, genetic stability, production of impurities or homologs, phenotypes, or shear resistance. After the initial primary screen, a more detailed secondary screen on approximately 10% of the most favorable clones is completed.

Culture preservation is vital to the success of a process development program. Reasons for establishing a strict cell banking procedure in development programs are similar to those motivating cGMP (current Good Manufacturing Practices) banking procedures and regulations imposed by the Food and Drug Administration (FDA) [5]. Effective preservation provides a long-term source of the cell line and a consistent initiation point for all development experiments [4, 6, 7]. Process development and strain improvement programs may capitalize on relatively small improvements [8]; without a consistent starting point such improvements may go unnoticed or perceived successes may not be reproducible. The cell bank is also important in determining the genetic stability of a new strain [9]. A number of different methods of preservation and the implications of each method have been documented in detail [6, 10–15]. The most effective and preferred of these methods involves the storage of cell lines in liquid nitrogen or preservation by freeze drying. As will be discussed later, the storage of cell lines on agar plates or slants is less favorable in that such storage frequently results in process variability [16]. The most suitable method of preservation, including the state at which cells are harvested for preservation (exponential growth, stationary phase, or as spores), the freezing medium, the method of preservation, and the recovery method [4, 6, 7] must be determined. Ultimately, the viability of the strain and the retention of production characteristics dictate the most suitable preservation method [14, 17].

6.2 Media Development

The objective in many media development programs is to improve the volumetric productivity (grams per liter per day) by evaluating the carbon, nitrogen, vitamin/growth factor, and inorganic nutrition requirements of the culture [4, 7, 18]. The solution to this aim will likely entail increasing the cell density while at the same time providing an environment conducive to maximizing the specific productivity ($1/X$ (dP/dt), Q_p). Although there exist a number of reports on improving secondary metabolite productivity [4, 7, 19, 20], media development remains an iterative process. This process can be systematically approached with the implementation of statistical media design [7, 21–23].

Several excellent references on microbial cell requirements and media development exist [4, 7, 18, 24, 25].

The focus of a media development program may be on completely defined media [26–29] and/or complex media [4, 7, 30]. A synthetic medium provides a better opportunity to monitor and define the nutritional requirements of a culture. However, productivity and cost are often superior with a complex medium, leading to its frequent use for manufacturing-scale production [30]. With a complex medium it is difficult to monitor nutrients, metabolites, and frequently cell density. Lot-to-lot variability of the composition of these undefined components also represents a real problem, both to the development scientist and the manufacturing plant manager. In addition, complex media may result in more difficult analytical and purification processes. With distinct benefits to each, both complex and defined media development avenues may be appropriate options to explore.

Although often overlooked, the identification of a suitable shake flask model is a requirement in any media development program [7, 31, 32]. It is important to understand the limitations of shake flask cultures, e.g., dissolved oxygen transfer, pH changes, and evaporation [7, 31]. As the media development program progresses, what once was not a problem may quickly become an issue. As an example, early in the development process the cell density supported by a given medium may not result in an oxygen consumption rate greater than the oxygen transfer rate. However, as media improvements are implemented, the cell density may become limited by the oxygen supplied in a given shake flask configuration. Variables to manipulate in defining an optimal shake flask model include the agitation rate and throw of the shaker, volume of media in a given flask, flask configuration, foam control, and nutrient concentration [7, 31, 32].

In performing media development, it is critical to monitor pH and identify a suitable buffer as early in the development process as possible. Although no buffer is ideal, the identification of a buffer that maintains the pH in an acceptable range is important to media improvement processes. Since polyketide production has been demonstrated to be sensitive to pH [33, 34], monitoring of the pH in flask cultures is required. In situations where changes in fermentation conditions result in significant pH changes, it may be necessary to change buffers as media changes are implemented.

Several studies have demonstrated the success of medium optimization in polyketide synthesis (for examples see [23, 25, 35–37]). In addition to improving the productivity of a culture, development studies have demonstrated that the distribution of polyketide products, both related and unrelated to the polyketide of interest, may be influenced by media development [38, 39].

6.3
Small-Scale Fermentation

Small-scale fermentation here refers to cultivation in bioreactors that can rigorously control the growth and production environment. These fermenters may range in size from 2 l to 20 l. There exist a handful of operating parameters that should be optimized near the beginning of the development process, including temperature (often done in shake flasks), pH, dissolved oxygen tension, and agitation rate, which affects both shear and dissolved oxygen [40]. All of these variables have been shown to affect culture growth and productivity in a variety of organisms [4, 19, 40, 41]. Although most polyketides are produced as secondary metabolites, i.e., products synthesized after the growth phase of a fermentation, this is not universally the case. For example, novel polyketides produced by heterologous hosts may demonstrate some degree of growth-associated production. In addition, media and culture conditions can be manipulated to result in production of secondary metabolites during the exponential growth phase. In cases where there is a degree of growth-associated production, the specific growth rate of an organism (μ) and its affect on the specific productivity (Qp) may be investigated using a chemostat.

Understanding what limits growth and the specific productivity in a medium may lead to the development of fed-batch processes that yield improved titers and volumetric productivity [26, 42–44], with the potential to minimize impurities [38]. Fed batch processes are sometimes controlled by off-line or on-line measurements, the preferred being on-line. Such processes are effective when nutrient or precursor feed rates are based on a sensitive and dependable measurement. For example, it is possible to maintain a limiting carbon supply by controlling the carbon source feed via the dissolved oxygen signal. The dissolved oxygen will decrease below a given set point when excess carbon is available, and then increase above the set point when carbon is limiting. Such a process has been demonstrated to scale well in a number of different production systems. In addition to being a suitable control parameter, dissolved oxygen has been demonstrated to have an important effect on the production of various polyketides [19, 37, 40]. Fed-batch, semi-continuous, and continuous processes provide a means to increase the volumetric productivity of a process [4, 24, 36, 41, 45].

The more analytical tools that are available and the better the understanding of critical biochemical pathways, the more rapidly fermentation processes can be developed. Besides those previously mentioned, a number of different parameters have been monitored on-line in fermentation development [7], including exhaust gas analysis and gas fluxes [46], cell density [47], redox potential [48], IR [49], culture fluorescence [50], biological activities [45], and viscosity. It is important to iterate that small-scale fermentation studies should aim to develop relatively simple control systems that are easily scaled. As an example, although HPLC systems are routinely set-up on line to measure and control laboratory scale fermentations, the robustness of such a system and its utility in a manufacturing facility remains debatable.

Scale-down studies are a valuable tool in fermentation development [51]. If production is going to occur in a fermentor for which the K_La or other parameter is precisely defined [52], correct down-scaled reactors should be used to mimic such configurations. Scale-down studies are helpful in that restrictions due to scale up are known in advance, thus minimizing small-scale studies that do not satisfy the ultimate good.

Besides optimizing environmental conditions, the inoculum procedure must be studied in detail [34, 53–56]. Inoculum procedure refers to the transfer of cell bank to growth medium and the steady expansion of a healthy culture until it is sufficient to inoculate the large-scale reactor [4, 57]. A consistent inoculum process is critical to a consistent manufacturing process. It is helpful to develop an inoculum procedure that does not require the use of agar slants or plates [58]. Transfer from plates to liquid has been demonstrated to be a source of variability [16]. Scale-down studies are also valuable in inoculum development. The additional passages a culture makes before a sufficient inoculum is available to inoculate a pilot or industrial scale fermenter can be accounted for using rigorous scale-down studies. Other aspects that must be addressed include the inoculum concentration and the medium used in the inoculum process.

7 Metabolic Engineering for the Overproduction of Polyketides

7.1 Heterologous Expression of Polyketides

Over the past few years, the expression of all or part of a polyketide pathway in a genetically friendly heterologous host is becoming an increasingly attractive alternative to performing strain and process improvement for polyketide production in the native producing organism. There are three key advantages of heterologous expression for the overproduction of a new polyketide metabolite. First, heterologous expression offers the advantage of PKS protein overexpression compared to native producing hosts, since well-developed promoter-regulator systems can be used. Second, by using a genetics-friendly and fast-growing heterologous host, it is possible to enhance the volumetric productivity of a fermentation process through a combination of random and directed approaches. Third, since polyketide biosynthesis is a relatively homogenous process (i.e., both the precursors and the enzymes are closely related for different polyketides), it is possible to re-use productive strategies for overproduction in different cases, and the process of polyketide overproduction does not have to be individualized for each product.

It should be noted that heterologous expression of a polyketide pathway itself does not lead to metabolite over-production. Indeed, titers are often below or at par with the natural host when the genes are first expressed in a heterologous host. However, the use of genetically and physiologically well-characterized hosts, as well as defined promoters and regulators, facilitates the improvement of the manufacture process at a more rapid pace.

The most well-established system for heterologous expression involves the hosts *S. coelicolor* or its close relative *S. lividans*, and a bifunctional actinomyces-*E. coli* vector with control elements for PKS gene expression that have been derived from the actinorhodin gene cluster [59]. This host-vector system has successfully been used to reconstitute functionally the polyketide pathways associated with biosynthesis of frenolicin [60], tetracenomycin [59], oxytetracycline [61], erythromycin [62], picromycin/methymycin [63], oleandomycin

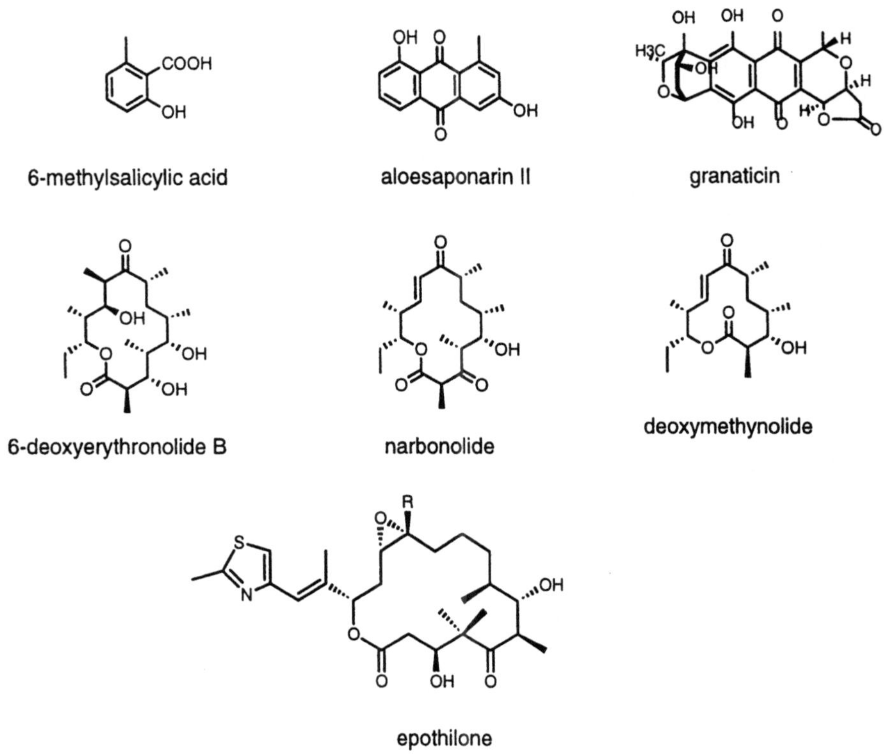

Fig. 2. Examples of heterologous polyketides that have been produced in *Streptomyces coelicolor*

[96], 6-methylsalicylic acid [64], and epothilone [65] (for examples, see Fig. 2]. The resulting polyketide products are typically generated in yields that may range between 1 mg/l and 100 mg/l culture. Moreover, PKS proteins are produced at 1–5% total cellular protein levels. Indeed, with the recent cloning and analysis of the enzymes responsible for post-translational modification of acyl carrier proteins (ACPs) [66], it has even become possible to express functional PKSs in *E. coli* [67]. While this is proving to be an excellent source for active protein preparations, the absence of specialized precursors such as methylmalonyl-CoA in *E. coli* has precluded the production of certain types of reporter metabolites in vivo. The successful expression of metabolically active levels of enzymes capable of in vivo synthesis of the correct isomer of methylmalonyl-CoA in *E. coli* has eliminated this limitation now [97].

7.2
Maximizing Gene Expression

Polyketides are typically produced by their host organism at the onset of stationary phase in response to various intracellular, intercellular, and external stimulating factors. The number of regulatory elements governing the expression of polyketide biosynthetic genes is quite large in actinomycetes - over a dozen genes related to production of actinorhodin have been identified in *S. coelicolor* [68]. These form complex, environmentally dependent regulatory networks which make it difficult to focus on a single regulatory cascade to engineer high expression levels. A more thorough picture should develop as genome-wide studies are undertaken (see below); however, the current set of known regulatory genes and expression tools offer a plethora of approaches to modulate empirically the expression levels of PKSs.

As many secondary metabolites can be produced by a single Actinomycete host organism, regulatory proteins are generally divided into two classes - pathway specific regulators which affect only a single polyketide (or other natural product) pathway and global (or pleiotropic) regulators which affect multiple or all pathways. The signaling pathways of the latter are less understood and are often coupled to physiological differentiation such as formation of aerial hyphae and spores. Many of these regulators belong to a unique family of transcriptional activators called *Streptomyces* antibiotic regulatory proteins (SARPs) [69]. The best-studied examples are the ActII-ORF4 and AfsR activators from *S. coelicolor* and the DnrI activator from *S. peucitius*. ActII-ORF4 and DnrI bind directly to promoters for the aromatic PKSs which produce actinorhodin and daunorubicin, respectively [70–72]. The ActII-ORF4/PactI activator-promoter system has been used to express many PKSs in *S. coelicolor* and *S. lividans* (see above). Overexpression of ActII-ORF4 has been shown to increase production of actinorhodin in *S. coelicolor* [73] and was also used to increase erythromycin production in a strain of *S. erythraea* [74]. Likewise, overexpression of DnrI led to overexpression of the daunorubicin PKS in *S. peucetius* [75]. AfsR, whose target is unknown, is a conditional global regulator which can increase actinorhodin when overexpressed in *S. coelicolor* under certain growth conditions.

Several other global regulators of actinorhodin biosynthesis have also been overexpressed or inactivated achieving a similar effect. However, several attempts to utilize many of the above genes to increase expression levels of the erythromycin modular PKS in *S. coelicolor* failed to provide any significant enhancements (R. McDaniel and P. Licari, unpublished observations). It is also surprising to find that, despite being one of the most intensely studied and successfully engineered PKS gene clusters with respect to polyketide biosynthesis, the erythromycin cluster remains one of the most poorly understood for regulation.

In addition to the natural control elements of polyketide production, many tools developed for heterologous expression of genes in *Streptomyces* may be of use for maximizing expression of PKS genes. For example, the tipA promoter [76], which is induced by addition of thiostrepton and the strong constitutive ermE* promoter [77] offer very different methods to control the timing and level of PKS expression. Heterologous expression of PKSs generally occurs either on low copy or chromosomal integrating vectors [59, 62, 78, 79]. Unfortunately, the use of high copy expression plasmids to increase PKS gene copy number has not been possible, which may either be due to plasmid instability or toxicity effects.

A final consideration for overexpression of PKSs is the post-translational phosphopantethienylation required to generate biosynthetically active PKS enzymes. This modification is performed by an enzyme called a holo-ACP synthase [66] (Fig. 3). Co-expression of a holo-ACP synthase with appropriate specificity is required for expression of PKSs in *E. coli* and yeast [67, 80]. Although most *Streptomyces* hosts appear to posses such enzymes with sufficiently relaxed specificities for heterologous expression, it may be necessary to concomitantly overexpress a holo-ACP synthase in *Streptomyces* strains which overexpress a PKS to obtain complete phosphopantethienylation. Related to this point is the mysterious thioesterase-like protein (TEII) which is found in many modular PKS gene clusters. Though the function of these proteins are unknown, inactivation of these enzymes by gene disruption generally leads to a tenfold or more decline in production of the corresponding polyketide metabolite [81–83]. As a result, many have speculated that TEIIs are involved in hydrolysis of aberrant thioesters bound to the PKS, which would otherwise block production. This has not been established, but if true, TEIIs may also require overexpression in some circumstances.

7.3 Enhancing Precursor Supplies

Overexpression of polyketide biosynthetic genes is probably not sufficient alone to achieve production levels near those of industrial developed strains. An appropriate flux of polyketide precursor substrates is also necessary. Labeling studies have shown that the acyl-CoA thioester building blocks used in polyketide biosynthesis may be derived from a variety of sources including carbohydrates, carboxylic acids, fatty acids, and amino acids (Fig. 4). Naturally, media composition is important and has a profound impact on polyketide production. However, media optimization can be limited by key enzymatic steps which represent bottlenecks in the conversion of these carbon sources to polyketide precursors. Metabolic pathways for polyketide precursors in *Streptomyces* are much like the regulatory pathways discussed above, complex and not well understood. Although most of the biosynthetic genes required for production of a polyketide are clustered within genomes, the genes for the most commonly used precursors – acetyl-CoA, propionyl-CoA, malonyl-CoA, and methylmalonyl-CoA – are distributed elsewhere in the genome, making it difficult to identify the most pertinent precursor pathways.

Since acetyl-CoA and malonyl-CoA are components of primary metabolism, it is generally assumed that these two substrates are in abundant supply. Therefore, most research has focused on engineering pathways to methylmalonyl-CoA and the more unusual precursors. At least two routes to methylmalonyl-CoA have been investigated in actinomycetes – carboxylation of propionyl-CoA and rearrangement of succinyl-CoA. A propionyl-CoA carboxylase gene cloned from the erythromycin producer *S. erythraea* did not significantly affect erythromycin production when inactivated, suggesting the latter pathway as the primary source of methylmalonyl-CoA in this organism [84]. However, this enzyme and its homologs from other sources [85] may be good candidates

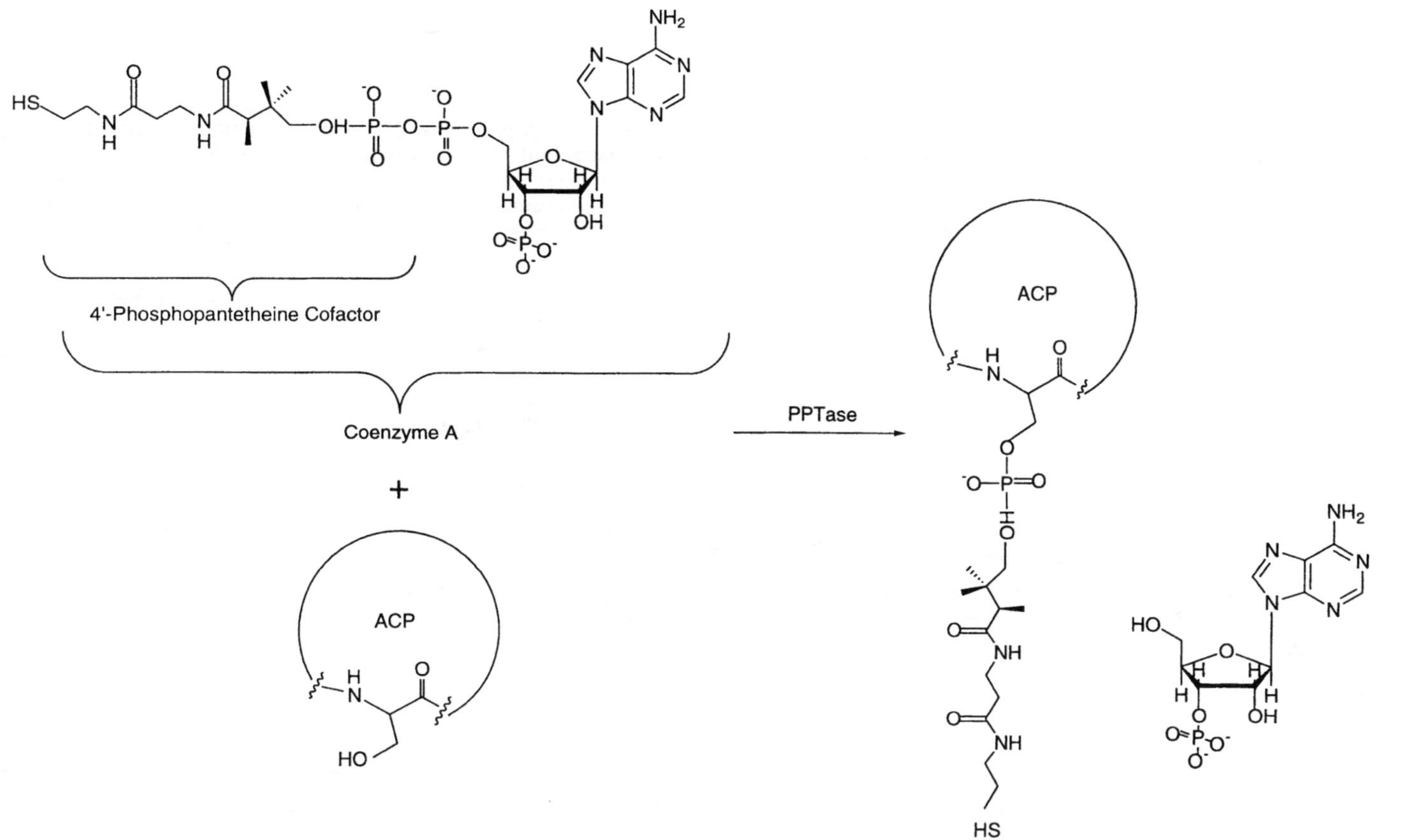

Fig. 3. Phosphopantetheinylation of the acyl carrier protein (ACP) domain of a polyketide synthase. In order to be active, polyketide synthases must be post-translationally modified by a family of enzymes called phosphopantetheine transferases (PPTases). These enzymes transfer the 4′-phosphopantetheine arm of Coenzyme A to an active site serine residue in the ACP

A

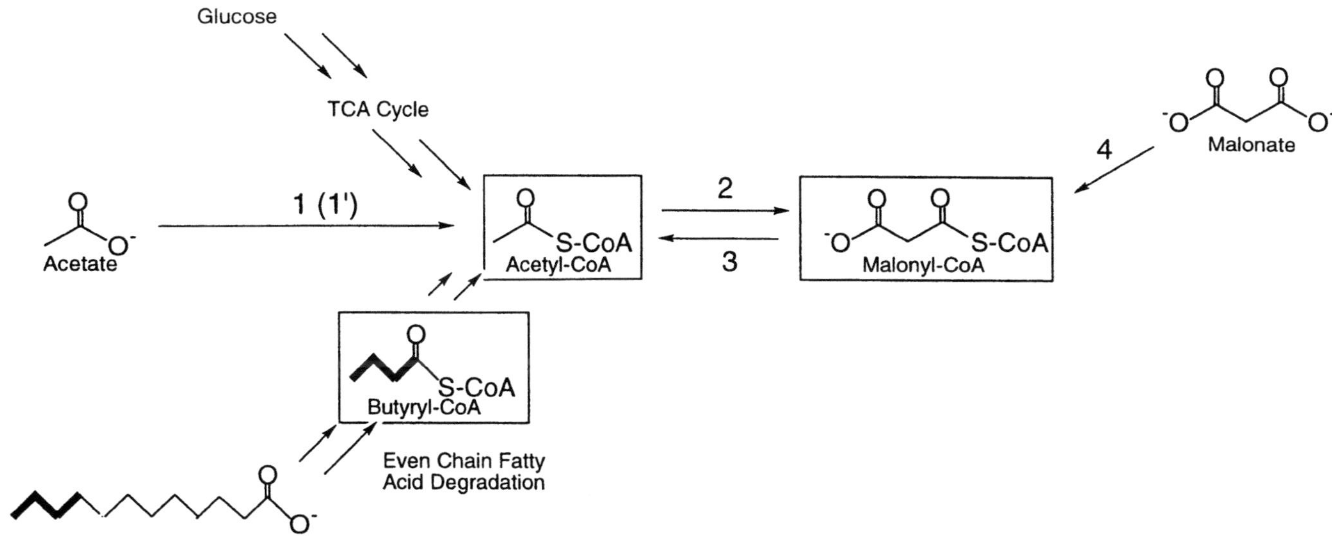

Fig. 4 A, B. Polyketide synthase substrate routes. Potential substrates have been boxed: **A** enzymes performing one enzymatic conversion: 1, acetyl-CoA synthetase (alternatively, 1′ represents a two enzyme pathway, acetate kinase followed by acetylphosphotransferase); 2, acetyl-CoA carboxylase; 3, malonyl-CoA decarboxylase; 4, malonyl-CoA synthetase; **B** enzymes performing one enzymatic conversion: 1, propionyl-CoA synthetase (1′, propionate kinase followed by propionylphosphotransferase); 2, propionyl-CoA carboxylase; 3, methylmalonyl-CoA decarboxylase; 4, methylmalonyl-CoA epimerase; 5, methylmalonyl-CoA mutase; 6, isobutyryl-CoA mutase

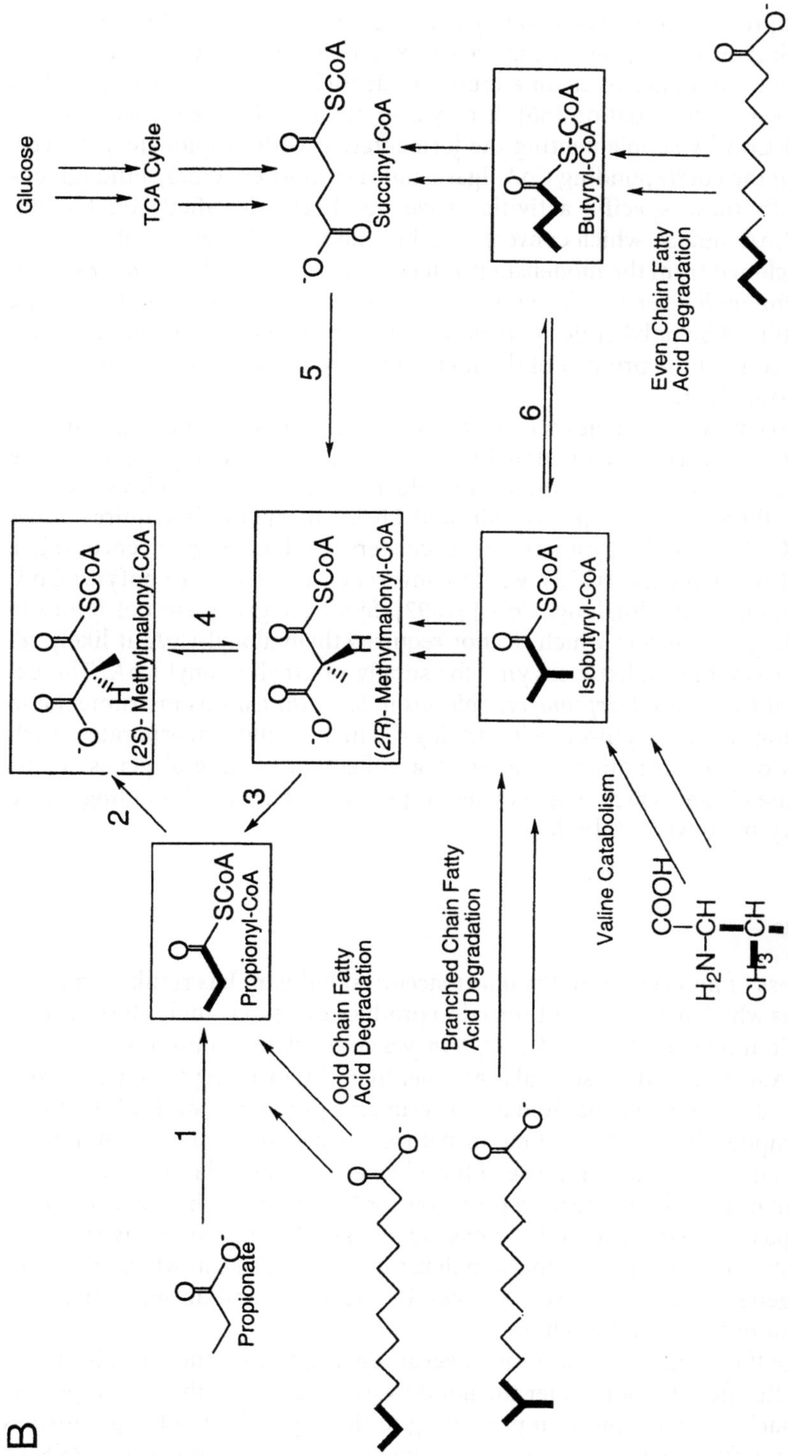

Fig. 4 A, B. (continued)

for overexpression to increase methylmalonyl-CoA availability. In this case, it becomes important to provide high levels of propionyl-CoA, which can be derived from both branched chain amino acid degradation and odd or branched chain fatty acid degradation [86]. It may also be possible to enhance levels of propionyl-CoA by supplementing the fermentation with propionate and overexpressing the corresponding CoA ligase and transport enzymes, although enzymes with these specific activities have not been identified yet. Methylmalonyl-CoA mutase, which converts succinyl-CoA to (*R*)-methylmalonyl-CoA has been cloned from the monensin producer, *Streptomyces cinnamonensis* [87, 88] and *Propionibacterium shermanii* [89]. Since (*S*)-methylmalonyl-CoA is the isomer utilized by polyketide synthases, coexpression with an epimerase is required to convert the product of the methylmalonyl-CoA mutase to the correct PKS substrate [90].

Clues to pathways or genes critical for the production of the more uncommon precursors, such as 2-ethylmalonyl-CoA, or 2-hydroxy-malonyl-CoA, can be found by gleaning the biosynthetic gene clusters for the polyketides which incorporate them. For example, tylosin and FK520 biosynthesis requires ethylmalonyl-CoA. Both biosynthetic gene clusters contain a gene encoding a crotonyl-CoA reductase (CCR), which converts crotonyl-CoA to butyryl-CoA, the precursor of ethylmalonyl-CoA [91, 92]. Since *ccr* genes are not found in polyketide gene clusters which do not require ethylmalonyl-CoA, it likely represents a key biosynthetic enzyme for supply of ethylmalonyl-CoA. The expression of CCR from *Streptomyces collinus* in *S. erythraea* was instrumental in engineering novel erythromycin analogs which could incorporate ethylmalonyl-CoA [93]. Another example is a collection of several genes in the FK520 gene cluster which is speculated to provide the unusual building block, 2-methoxy-malonyl-CoA [92].

7.4
Superhosts

The success of polyketides in the pharmaceutical industry has resulted in many organisms which have been optimized to produce extremely high titers of compound. Commercial strains of *Streptomcyes* and related actinomycetes exist that, for example, produce several grams per liter of tylosin, erythromycin, avermectin, and oxytetracycline. To date, experiments performed on PKSs to create novel compounds have been done in naturally isolated strains which are low producers in comparison. Because of the advances in molecular biology and development of genetic engineering protocols for *Streptomyces* that have occurred over the past 20 years, it should be possible to take advantage of industrially developed strains to engineer generic polyketide 'superhosts', in which heterologous or genetically engineered PKSs can be expressed and an immediate improvement in titers can be achieved.

Despite the number of polyketide overproducing strains, there has been relatively little effort to look under the hood to see what drives these high performance machines. One question pertaining to the applicability of superhosts is whether the titer increases result from catalytic improvements of the PKS, or from background effects such as increased expression levels or precursor supply. This was recently addressed by examining the activity of 6-deoxyerythronolide B synthase (DEBS) isolated from an overproducing *S. erythraea* strain. The production levels of DEBS from this strain and DEBS isolated from the wild-type strain were similar when expressed in a non-overproducing host (R. McDaniel, unpublished). This suggests that the factors contributing to overproduction lie predominantly in expression levels and/or precursor availability and that PKS optimization may not be important. Therefore, the possibility that heterologous PKSs will overproduce when expressed in a superhost is promising and efforts are underway to test this hypothesis.

Superhosts also possess the ability to increase the size of 'unnatural' natural product libraries that are generated from genetically engineered PKSs. The overproducing erythromycin strain discussed above was also used to determine that production levels from genetically engineered PKSs could be enhanced significantly in the overproduction background of this host. A 100-fold improvement in titer from a genetically modified DEBS was obtained (R. McDaniel, unpublished). Because the production levels from genetically engineered PKSs generally correlates to the number of modifications that have been made (i.e., the more changes, the lower the production levels) [94], increasing the basal level of polyketide should allow more permutations to be introduced before production becomes too low.

7.5 Genomics Guided Process and Strain Improvement

Although generally touted as a significant advancement for basic biology and drug development, the field of genomics also has great potential to aid metabolic engineering and strain improvement. With the completion of the *S. coelicolor* genome sequencing project due for completion by the end of year 2000 and the number of genomics tools that currently exist and are being developed at a rapid pace, the actinomycete community is now poised to develop a better picture of the complex metabolic pathways in these organisms and how they are affected in various environments. These tools can be used to learn what differentiates a low producer from an overproducer and, in turn, used to short-cut traditional strain improvement by deleting or overexpressing corresponding genes. Further improvements to existing overproducing strains may also be observed because mutations can be engineered which are difficult to access by random mutagenesis. One drawback often encountered with industrial production strains is a high barrier to genetic manipulation, which is not understood. Finally, genomics tools can also be used to obtain a comprehensive readout on cellular states in fermentation conditions. This allows the process engineer to correlate good and bad effects on production levels to molecular pathways in the cell, providing a more rational and direct approach to optimizing fermentation parameters. Some detailed examples of how genomic technologies can be used for process and strain improvement have been described [95].

8 Conclusion

There are only a few thousand polyketide natural products that have been discovered from microorganisms to date. Dozens of these have been developed as commercial products. With the elucidation of molecular mechanisms that underlie polyketide biosynthesis, it has become possible to exploit better the pharmacological properties of known polyketides, and also to create new bioactive compounds via genetic engineering. These developments have prompted a reappraisal of established practices in bioprocess engineering for polyketide production, and have catalyzed the emergence of new metabolic engineering strategies for this purpose. Given the enormous successes of coordinated genetic engineering and process engineering in the recombinant biopharmaceutical industry, one could foresee similar efforts having a comparable impact on the future utility of polyketides to mankind.

References

1. Cane DE, Walsh CT, Khosla C (1998) Harnessing the biosynthetic code. Combinations, permutations, mutations. Science 282:63–68
2. Khosla C, Gokhale R, Jacobsen JR, Cane DE (1999) Tolerance and specificity of polyketide synthases. Ann Rev Biochem 68:219–253
3. Strohl WR (1997) Biotechnology of antibiotics. Marcel Dekker, Inc
4. Stanbury PF, Whitaker A (1986) Principles of fermentation technology. Pergamon Press, Oxford, UK
5. Steuer A, Ostrove JM, McEntire J (1996) Establishing cell banks under current good manufacturing practices. Biopharm 9:40–44
6. Brown MRW, Gilbert P (1995) Microbiological quality assurance screening and bioassay: a guide towards relevance and reproducibility of inocula. CRC Press, Boca Raton, FL
7. Demain AL, Davies JE (1999) Manual of industrial microbiology and biotechnology. American Society for Microbiology, Washington, DC
8. Davies OL (1964) Screening for improved mutants in antibiotic research. Biometrics Sept:576–591
9. Roth M, Noack D (1982) Genetic stability of differentiated functions in *Streptomyces hygroscopicus* in relation to conditions in continuous culture. J Gen Microbiol 128:107–114
10. Berny JF, Hennebert GL (1991) Viability and stability of yeast cells and filamentous fungus spores during freeze drying: effects of protectants and cooling rates. Mycologia 83:805–815
11. Chen YY (1987) The preservation of basidiomycetes cultures by ultra-low temperature freezing. Acta Mycol Sin 6:110–117
12. Gerhardt P, Murray RG, Gstilow RN, Nester EW, Wood WA (1981) Manual of methods for general bacteriology. American Society for Microbiology, Washington, DC
13. Kirsop BE, Snell JJ (1985) Maintenance of microorganisms. Academic Press, London
14. Lang E, Malik KA (1996) Maintenance of biodegradation capacities of aerobic bacteria during long term preservation. Biodegradation 7:65–71
15. Nabais AMA, da Fonseca MMR (1995) The effect of solid medium composition on growth and sporulation of *Streptomyces clavuligerus*: spore viability during storage at 4°C. Biotechnol Tech 9:361–364
16. Bloomfeld SF, Arthur M, Gibson H, Morley K, Gilbert P, Brown MRW (1995) Development of reproducible test inocula for disinfectant testing. Int Biodeterior Biodegrad 36:311–331

17. Monaghan RL, Currie SA (1985) Preservation of antibiotic production by representative bacteria and fungi. Dev Ind Microbiol 26:787–792
18. Atkinson B, Mavituna, F (1983) Stoichiometric aspects of microbial metabolism. In: Biochemical engineering and biotechnology handbook. MacMillan, London, pp 114–203
19. Clark GJ, Langley D, Bushnell ME (1995) Oxygen limitation can induce microbial secondary metabolite formation: investigations with miniature electrodes in shaker and bioreactor culture. Microbiology 141:663–669
20. Doull JL, Vining LC (1990) Nutritional control of actinorhodin production by *Streptomyces coelicolor* A(3) 2: suppressive effects of nitrogen and phosphate. Appl Microbiol Biotechnol 32:449–454
21. Chen HC (1996) Optimizing the concentrations of carbon, nitrogen, and phosphorus in a citric acid fermentation with response surface method. Food Biotechnol 10:13–27
22. Haaland PD (1989) Experimental design in biotechnology. Marcel Dekker, New York,NY
23. Saval S, Pablos L, Sanchez S (1993) Optimization of a culture medium for streptomycin production using response-surface methodology. Bioresour Technol 43:19–25
24. Pirt SJ (1975) Principles of microbe and cell cultivation.Wiley, New York
25. Strobel RJ, Nakatsukasa WM (1993) Response surface methods for optimizing *Saccharopolyspora spinosa*, a novel macrolide producer. J Ind Microbiol 11:121–127
26. Boeck LD, Favret ME, Wetzel RW (1992) Biosynthesis of thiopeptide antibiotic A10255 in stirred reactors using a chemically defined medium supplemented with continuous nutrient feeds. J Antibiot 45:1278–1285
27. Hosler P, Johnson MJ (1953) Penicillin from chemically defined media. Ind Eng Chem 45:871–874
28. Jain D, Nielson JBK, Buckland BC (1992) Kinetics of efrotomycin synthesis in a synthetic fermentation medium. Bioprocess Eng 7:257–263
29. Zhang J, Marcin C, Shifflet MA, Salmon P, Brix T, Greasham R, Buckland BC, Chartrain M (1996) Development of a defined medium fermentation process for physostigmine production by *Streptomyces griseofuscus*. Appl Microbiol Biotechnol 44:568–575
30. Zabriskie DW, Armiger WB, Phillips DH, Albano PA (1994) Traders guide to fermentation media formulation. Trader's Protein, Memphis, TN
31. Henzler HJ, Schedel M (1991) Suitability of the shaking flask for oxygen supply to microbiological cultures. Bioprocess Eng 7:123–131
32. McDaniel LE, Bailey EG (1969) Effect of shaking speed and type of closure on shake flask cultures. Appl Microbiol 17:286–290
33. Espeso EA, Tilburn J, Arst HN, Penalva MA (1993) pH regulation is a major determinant in expression of a fungal penicillin biosynthetic gene. EMBO J 12:3947–3956
34. Glazebrook MA, Vining LC, White RL (1992) Growth morphology of *Streptomyces akiyoshiensis* in submerged culture: influence of pH, inoculum and nutrients. Can J Microbiol 38:98–103
35. McDaniel LE, Bailey EG, Ethiraj S, Andrews HR (1976) Application of response surface optimization techniques to polyene macrolide fermentation studies in shake flasks. Dev Ind Microbiol 17:286–290
36. McIntyre JJ, Bull AT, Bunch AW (1996) Vancomycin production in batch and continuous culture. Biotechnol Bioeng 26:916–925
37. Uchida M, Sawada H, Asai T, Suzuki M (1981) Effect of inorganic phosphate and dissolved oxygen on production of maridomycin. J Ferment Technol 59:399–401
38. Junker B, Reddy J, Olewinski R, Gailliot P, Byrne K, Gbewonyo K (1998) Secondary metabolite scale-up to minimize homolog impurity levels. Biotechnol Bioeng 59:595–604
39. Adamcova J, Proksa B, Fuska J (1992) Regulation of biosynthesis of vermiculin and vermistatin in *Penicillium vermiculatum*. Folia Microbiol 37:50–52
40. Heyderian SM, Lilly MD, Ison AP (1996) The effect of culture conditions on the production of erythromycin by *Saccharopolyspora erythraea* in batch culture. Biotechnol Lett 18:1181–1186

41. Fazeli MR, Cove JH, Baumberg S (1995) Physiological factors affecting streptomycin production by *Streptomyces griseus* 12475 in batch and continuous culture. FEMS Microbiol Lett 126:55–62
42. Ates S, Elibol M, Mavituna, F (1997) Production of actinorhodin by *Streptomyces coelicolor* in batch and fed-batch processes. Process Biochem 32:273–278
43. Gbewonyo G, Buckland BC, Lilly MD (1991) Development of a large-scale continuous substrate feed process for the biotransformation of simvastatin by *Nocardia* sp. Biotechnol Bioeng 37:1101–1107
44. Matelova V, Brecka A, Matouskova J (1972) New method of intermittent feeding in penicillin biosynthesis. J Appl Microbiol 23:669–670
45. Meyer C, Beyeler W (1984) Control strategies for continuous bioprocesses based on biological activities. Biotechnol Bioeng 26:916–925
46. Buckland BC, Brix T, Fastert H, Gbewonyo K, Hunt G, Jain D (1985) Fermentation exhaust gas analysis using mass spectrometry. Bio/Technology 3:982–988
47. Iijima S, Yamashita S, Matsunaga K, Miura H, Morikawa M, Shimizu K, Matsubara M, Kobayashi T (1987) Use of a novel turbidometer to monitor microbial growth and control glucose concentration. J Chem Technol Biotechnol 40:203–213
48. Kjaergaard L (1977) The redox potential: its use and control in biotechnology. Adv Biochem Eng 7:131–150
49. Rhiel M, Cannizzaro C, Marison I, von Stockar U (2000) All-in-one bioprocess monitoring with a mid-IR spectroscopic sensor. Abstracts of the 219th ACS National Meeting, San Francisco, March 2000
50. Humphrey A (1988) The potential of on-line fluorometric measurements for the monitoring and control of fermentation systems. Aust J Biotechnol 2:141–147
51. Jem J (1989) Scale-down techniques for fermentation. Biopharm March 30:39
52. Closterhuis NMG, Kossen NWE (1983) Dissolved oxygen concentration profiles in a production-scale bioreactor. Biotechnol Bioeng 26:546–550
53. Friedrich J, Cimerman A, Steiner W (1990) Production of pectolytic enzymes by *Aspergillus niger*: effect of inoculum size and potassium hexacyanoferrate II trihydrate. Appl Microbiol Biotechnol 33:377–381
54. Losey VA, Chagrin BA, Vadysheva TN, Shurygina LV (1988) Influence of seed material quality on cephalosporin C biosynthesis. Antibiot Khimioter 33:493–496
55. Warr SR, Gershater CJ, Box SJ (1996) Seed stage development for improved fermentation performance: increased milbemycin production by *Streptomyces hygroscopicus*. J Ind Microbiol 16:295–300
56. Webb C, Kamat SP (1993) Improving fermentation consistency through better inoculum preparation. World J Microbiol Biotechnol 9:308–312
57. Hopf NW, Yonsel S, Decker WD (1990) Ambruticin S. production in amino acid rich media. Appl Microbiol Biotechnol 32:499–504
58. Acreman J (1994) Algae and cyanobacteria: isolation, culture and long-term maintenance. J Ind Microbiol 13:193–194
59. McDaniel R, Ebert-Khosla S, Hopwood D, Khosla C (1993) Engineered biosynthesis of novel polyketides. Science 262:1546–1557
60. McDaniel R, Ebert-Khosla S, Hopwood DA, Khosla C (1993) Engineered biosynthesis of novel polyketides: manipulation and analysis of an aromatic polyketide synthase with unproven catalytic specificities. J Am Chem Soc 115:11,671–11,675
61. Fu H, Ebert-Khosla S, Hopwood DA, Khosla C (1994) Relaxed specificity of the oxytetracycline polyketide synthase for an acetate primer in the absence of a malonamyl primer. J Am Chem Soc 116:6443–6444
62. Kao CM, Katz L, Khosla C (1994) Engineered biosynthesis of a complete macrolactone in a heterologous host. Science 265:509–512
63. Tang L, Fu H, Betlach M, McDaniel R (1999) Elucidating the mechanism of chain termination switching in the picromycin/methymycin polyketide synthase. Chem Biol 6: 553–558

64. Bedford DJ, Schweizer E, Hopwood DA, Khosla C (1995) Expression of a functional fungal polyketide synthase in the bacterium *Streptomyces coelicolor* A3(2). J Bacteriol 177:4544–4548
65. Tang L, Shah S, Chung L, Carney J, Katz L, Khosla C, Julien B (2000) Cloning and heterologous expression of the epothilone gene cluster. Science 287:640–642
66. Lambalot RH, Gehring AM, Flugel RS, Lacelle M, Marahiel MA, Reid R, Khosla C, Walsh CT (1996) A new enzyme superfamily – the phosphopantetheinyl transferases. Chem & Biol 3:923–936
67. Gokhale RS, Tsuji SY, Cane DE, Khosla C (1999) Dissecting and exploiting intermodular communication in polyketide synthases. Science 284:482–485
68. Bibb M (1996) The regulation of antibiotic production in *Streptomyces coelicolor* A3(2). Microbiology 142:1335–1344
69. Wietzorrek A, Bibb M (1997) A novel family of proteins that regulates antibiotic production in streptomycetes appears to contain an OmpR-like DNA-binding fold. Mol Microbiol 25:1177–1184
70. Gramajo HC, Takano E, Bibb MJ (1993) Stationary-phase production of the antibiotic actinorhodin in *Streptomyces coelicolor* A3(2) is transcriptionally regulated. Mol Microbiol 7:837–845
71. Arias P, Fernandez-Moreno M, Malpartida F (1999) Characterization of the pathway-specific positive transcriptional regulator for actinorhodin biosynthesis in *Streptomyces coelicolor* A3(2) as a DNA-binding protein. J Bacteriol 181:6958–6968
72. Tang L, Grimm A, Zhang Y-X, Hutchinson CR (1996) Purification and characterization of the DNA-binding protein DnrI, a transcriptional factor of daunorubicin biosynthesis in *Streptomyces peucetius*. Mol Microbiol 22:801–813
73. Passatino R, Puglia A-M, Chater K (1991) Additional copies of the *act*II regulatory gene induce actinorhodin production in pleiotropic *bld* mutants of *Streptomyces coelicolor* A3(2). J Gen Microbiol 137:2059–2064
74. Rowe CJ, Cortes J, Gaisser S, Staunton J, Leadlay PF (1998) Construction of new vectors for high-level expression in actinomycetes. Gene 216:215–223
75. Stutzman-Engwall KJ, Otten S, Hutchinson CR (1992) Regulation of secondary metabolism in *Streptomyces* spp. and overproduction of daunorubicin in *Streptomyces peucetius*. J Bacteriol 174:144–154
76. Murakami T, Holt TG, Thompson CJ (1989) Thiostrepton-induced gene expression in *Streptomyces lividans*. J Bacteriol 171:1459–1466
77. Bibb MJ, White J, Ward JM, Jansenn GR (1994) The mRNA for the 23S rRNA methylase encoded by the ermE gene of *Saccaropolyspora erythraea* is translated in the absence of a conventional ribosome-binding site. Mol Microbiol 14:533–545
78. Ziermann R, Betlach M (2000) A two-vector system for the production of recombinant polyketides in *Streptomyces*. J Ind Microbiol Biotech 24:46–50
79. Xue Q, Ashley G, Hutchinson CR, Santi DV (1999) A multi-plasmid approach to preparing large libraries of polyketides. Proc Natl Acad Sci USA 96:11,740–11,745
80. Kealey JT, Liu L, Santi DV, Betlach MC, Barr PJ (1998) Production of a polyketide natural product in nonpolyketide-producing prokaryotic and eukaryotic hosts. Proc Natl Acad Sci USA 95:505–509
81. Xue Y, Zhao L, Liu H-W, Sherman DH (1998) A gene cluster for the macrolide antibiotic biosynthesis in *Streptomyces venezuelae*: architecture of metabolic diversity. Proc Natl Acad Sci USA 95:12,111–12 116
82. Butler AR, Bate N, Cundliffe E (1999) Impact of thioesterase activity on tylosin biosynthesis in *Streptomyces fradiae*. Chem & Biol 6:287–292
83. Reeves AR, English SR, Lampel JS, Post DA, Vanden Boom TJ (1999) Transcriptional organization of the erythromycin biosynthetic gene cluster of *Saccharopolyspora erythraea*. J Bacteriol 181:7098–7106
84. Donadio S, Staver MJ, Katz L (1996) Erythromycin production in *Saccharopolyspora erythraea* does not require a functional propionyl-CoA carboxylase. Mol Microbiol 19: 977–984

85. Rodriguez E, Gramajo H (1999) Genetic and biochemical characterization of the α and β components of a propionyl-CoA carboxylase complex of *Streptomyces coelicolor* A3(2). Microbiol 145:3109–3119
86. Tang L, Xhang Y-X, Hutchinson CR (1994) Amino acid catabolism and antibiotic synthesis: valine is a source of precursors for macrolide iosynthesis in *Streptomyces ambofaciens* and *Streptomyces fradiae*. J Bacteriol 176:6107–6119
87. Birch A, Leiser A, Robinson JA (1993) Cloning, sequencing, expression of the gene encoding methylmalonyl-coenzyme A mutase from *Streptomyces cinnamonensis*. J Bacteriol 175:3511–3519
88. Vrilbloed JW, Zerbe-Burkhardt K, Ratnatilleke A, Grubelnik-Leiser A, Robinson JA (1999) Insertional inactivation of methylmalonyl coenzyme A (CoA) mutase and isobutyryl-CoA mutase genes in *Streptomyces cinnamonensis*: influence on polyketide antibiotic biosynthesis. J Bacteriol 181:5600–5605
89. Marsh EN, Leadlay PF (1989) Methylmalonyl-CoA mutase from *Propionibacterium shermanii*. Evidence for the presence of two masked cysteine residues. Biochem J 260: 339–343
90. Marsden AFA, Caffrey P, Aparicio JF, Loughran MS, Staunton J, Leadlay PF (1994) Stereospecific acyl transfers on the erythromycin-producing polyketide synthase. Science 263:378–380
91. Merson-Davies LA, Cundliffe E (1994) Analysis of five tylosin biosynthetic genes from the *tylIBA* region of the *Streptomyces fradiae* genome. Mol Microbiol 13:349–355
92. Wu K, Chung L, Revill WP, Katz L, Reeves CD (2000) The FK520 gene cluster of *Streptomyces hygroscopicus* var. *ascomyceticus* (ATCC 14891) contains genes for biosynthesis of unusual polyketide extender units. Gene 251:81–90
93. Stassi DL, Kakavas SJ, Reynolds KA, Gunawardana G, Swanson S, Zeidner D, Jackson M, Liu H, Buko A, Katz L (1998) Ethyl-substituted erythromycin derivatives produced by directed metabolic engineering. Proc Natl Acad Sci USA 95:7305–7309
94. McDaniel R, Thamchaipenet A, Gustafsson C, Fu H, Betlach M, Betlach M, Ashley G (1999) Multiple genetic modifications of the erythromycin gene cluster to produce a library of novel "unnatural" natural products. Proc Natl Acad Sci USA 96:1846–1851
95. Kao CM (1999) Functional genomic technologies: creating new paradigms for fundamental and applied biology. Biotechnol Prog 15:304–311
96. Shah S, Xue Q, Tang L, Carney JR, Betlach M, McDaniel R (2000) Cloning, characterization and heterologous expression of a polyketide synthase and P-450 oxidase involved in the biosynthesis of the antibiotic oleandomycin. J Antibiotic 53:502–508
97. Pfeifer B, Admiraal SJ, Gramajo H, Care DE, Khosla C (2001) Biosynthesis of complex polyketides in a metabolically engineered strain of E. coli. Science 291:1790–1792

Received: November 2000

Metabolic Engineering of *Saccharomyces cerevisiae* for Xylose Utilization

Bärbel Hahn-Hägerdal[1], C. Fredrik Wahlbom[1], Márk Gárdonyi[1],
Willem H. van Zyl[2], Ricardo R. Cordero Otero[2], Leif J. Jönsson[1]

[1] Department of Applied Microbiology, Lund University, PO Box 124, 221 00 Lund, Sweden, e-mail: Barbel.Hahn-Hagerdal@tmb.lth.se

[2] Department of Microbiology, University of Stellenbosch, Private Bag XI, 7600 Stellenbosch, South Africa

Metabolic engineering of *Saccharomyces cerevisiae* for ethanolic fermentation of xylose is summarized with emphasis on progress made during the last decade. Advances in xylose transport, initial xylose metabolism, selection of host strains, transformation and classical breeding techniques applied to industrial polyploid strains as well as modeling of xylose metabolism are discussed. The production and composition of the substrates – lignocellulosic hydrolysates – is briefly summarized. In a future outlook iterative strategies involving the techniques of classical breeding, quantitative physiology, proteomics, DNA micro arrays, and genetic engineering are proposed for the development of efficient xylose-fermenting recombinant strains of *S. cerevisiae*.

Keywords. Xylose, *Saccharomyces cerevisiae*, Ethanol, Lignocellulose, Metabolic engineering

1	**Introduction**	54
1.1	Ethanol Production from Xylose	54
1.2	Abundance of Xylans in Lignocellulosic Raw Materials	55
1.3	Xylose Metabolism	55
2	**Substrate Composition**	57
2.1	Composition of Hemicellulose Hydrolysates	57
2.2	Hydrolysis of Lignocellulose Polysaccharides	59
2.3	Fermentation Inhibitors in Lignocellulose Hydrolysates	60
3	**Xylose Transport**	61
3.1	Xylose Transport in *S. cerevisiae*	62
3.2	Xylose Transport in Natural Xylose-Utilizing Yeasts	63
3.3	Engineering Xylose Transport in *S. cerevisiae*	64
4	**The Conversion of Xylose to Xylulose**	65
4.1	Xylose Reductase (XR)/Xylitol Dehydrogenase (XDH)	65
4.1.1	Activity Ratios for XR and XDH	65
4.1.2	Protein Engineering – Fusion Protein	66
4.1.3	Protein Engineering – Site-Specific Mutagenesis	66
4.1.4	Xylitol Transport	67
4.1.5	Oxygen Utilization	68
4.2	Xylose Isomerase	68

Advances in Biochemical Engineering/
Biotechnology, Vol. 73
Managing Editor: Th. Scheper

4.2.1 Expression of Xylose Isomerase in *S. cerevisiae* 68
4.2.2 Recent Developments to Improve the XI Activity 70

5 **Hexose and Pentose Fermentation** 70

6 **Choice of Host Strain** 72

7 **Metabolic Engineering of Polyploid Strains** 73

8 **Classical Breeding Techniques** 74

9 **Metabolic Modeling** 75

10 **Future Outlook** 78

References 79

1 Introduction

The present review summarizes the past decade's work on metabolic engineering of *Saccharomyces cerevisiae* to generate an efficient xylose-fermenting yeast. Metabolic engineering of yeasts may be carried out by recombinant DNA techniques, by clonal selection after mutagenesis, protoplast fusion, and hybridization. A combination of some or all of these methods may be ideal for the development of yeasts with novel metabolic activities. The review will especially highlight work on (i) xylose transport, (ii) xylose to xylulose conversion, (iii) differences between hexose and pentose fermentation, (iv) host strain selection, (v) transformation of industrial polyploid strains, (vi) classical breeding techniques, and (vii) modeling of xylose metabolism. Additionally, the composition of the fermentation substrate – the lignocellulose hydrolysate – is reviewed in relation to the origin of the lignocellulosic biomass and the physical, chemical and biochemical processes utilized to generate the monosaccharide substrate. Finally, in a future outlook, the results of metabolic engineering of *S. cerevisiae* are compared with other natural and recombinant xylose-fermenting microorganisms with a discussion of the pros and cons of different strain development strategies. While work on this review was in progress, two reviews in this field have been published [1, 2].

1.1 Ethanol Production from Xylose

When, in 1973, the OPEC countries reduced their oil production, it triggered an energy crisis worldwide and initiated research into and development of "energy from renewable resources". This included the bioconversion of agricultural and forest products into liquid transportation fuels such as ethanol [3, 4]. When restrictions on oil production were relieved, research and development on the bioconversion of renewable resources continued. One reason for this was that the consumption of transportation fuel worldwide continued to increase [5], the other reason being the recent awareness of possible global warming as a result of increased burning of fossil fuels [6–9]. Although the replacement of petrol by ethanol from renewable resources may be governed by environmental concerns, the production of ethanol fuel must be economically competitive to constitute a sustainable alternative. High product yield is an important criterion for all industrial processes. For fuel ethanol production it is crucial since the product has a low value and the raw material constitutes a major part of the production cost [10–12].

1.2
Abundance of Xylans in Lignocellulosic Raw Materials

Hemicellulose is one of the major components of lignocellulose. Depending on the nature of the raw material, the hemicellulose fraction contains varying levels of xylose-based hemicelluloses, xylans (Table 1). The xylan content is generally high in hardwood (wood from deciduous trees) and in agricultural residues, and somewhat lower in softwood (wood from coniferous trees).

As a result, a substantial fraction of the monosaccharides in lignocellulose hydrolysates from hardwood and agricultural residues consists of xylose. Consequently, ethanolic fermentation of xylose is of major concern for the efficient utilization of lignocellulosic hydrolysates to produce fuel ethanol.

1.3
Xylose Metabolism

Xylose can be fermented to ethanol by bacteria, yeast, and filamentous fungi (for reviews see [13–19]). In bacteria, the initial step in xylose metabolism is isomerization to xylulose. The enzyme xylose isomerase (XI) converts xylose to xylulose (Fig. 1). Certain bacteria can ferment all sugars in a lignocellulose hydrolysate, but produce a mixture of acids and solvents. *Escherichia coli* and *Klebsiella oxytoca* have been metabolically engineered to produce ethanol exclusively [20]. The bacterium *Zymomonas mobilis* can only utilize sucrose, glucose, and fructose, but ferments them to ethanol with yields equivalent to those obtained with yeast [21]. *Z. mobilis* has been metabolically engineered with a xylose-utilizing pathway [22] and an arabinose-utilizing pathway [23].

Table 1. Xylan content in wood and agricultural residues

Raw material	Xylan (% DW)	Reference
Hardwood	15–30	[59]
Softwood	5–10	[59]
Wheat straw	19	[57]
Sugarcane bagasse	21	[57]
Corn fiber	37	[60]

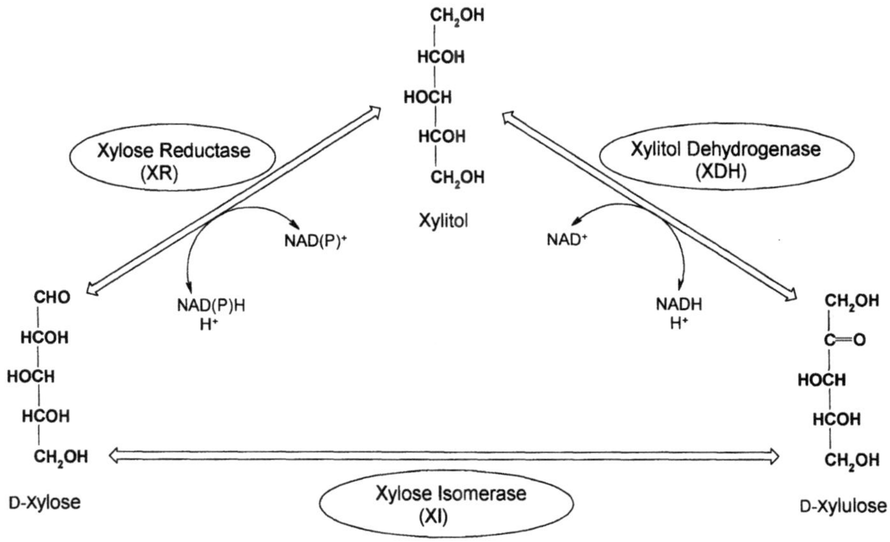

Fig. 1. The interconversions between xylose, xylitol, and xylulose

The resulting recombinant strains produce ethanol from xylose and arabinose.

In yeast and filamentous fungi the initial steps of xylose metabolism involve reduction to xylitol by the enzyme xylose reductase (XR), followed by the oxidation of xylitol to xylulose by the enzyme xylitol dehydrogenase (XDH) [24–27]. Although filamentous fungi generate ethanol concentrations and yields comparable to those obtained in hexose fermentation with yeast, the productivity is too low to be economically feasible [28–30]. The yeasts *Pichia stipitis* [31], *Candida shehatae* [31], and *Pachysolen tannophilus* [32] have been singled out as efficient fermenters of xylose to ethanol. They require a low and well-controlled level of aeration for maximal ethanol production [33]. *P. stipitis* and *C. shehatae* are extremely sensitive to metabolic inhibitors present in lignocellulose hydrolysates [34, 35], whereas *P. tannophilus* converts a major fraction of the xylose substrate to xylitol [34].

In industrial ethanol fermentation processes the preferred organism is the yeast *S. cerevisiae*. It produces ethanol from hexoses in industrial, non-sterilized raw materials, such as molasses with product concentrations of approximately 50 g l^{-1}, and product yields of 0.5 g g^{-1} with productivities of 2 g l^{-1} h^{-1} [36]. *S. cerevisiae* carries genes encoding an unspecific aldose reductase with XR activity [37, 38] and an unspecific sugar alcohol dehydrogenase with XDH activity [39], but cannot convert xylose to ethanol, only the isomer xylulose [40–44]. The *P. stipitis* genes *XYL1* and *XYL2* encoding XR and XDH, respectively, have been actively expressed in *S. cerevisiae*, which generated xylose utilizing recombinant strains [45–47]. These strains produced xylitol from xylose rather than ethanol. It was suggested that the endogenous activities of the enzymes xylulokinase (XK) [48] and transaldolase (TAL) [49, 50] imposed limitations on the yeast *S. cerevisiae* during xylose utilization. Numerous attempts to construct xylose-fermenting *S. cerevisiae* strains by introducing the bacterial gene *xylA* encoding xylose isomerase (XI) have failed [51–55]. Ethanol formation from xylose has only been demonstrated when the *xylA* gene from the thermophilic bacterium *Thermus thermophilus* was expressed in a recombinant strain of *S. cerevisiae* [56].

In the following sections the fermentation of xylose to ethanol by recombinant strains of *S. cerevisiae* will be discussed in relation to the absence of specific xylose transporters in *S. cerevisiae*, and the strategies for co-factor balancing in strains expressing *XYL1* and *XYL2*. Differences between hexose and pentose fermentation will be highlighted. The possibility of preselecting host strains with desirable qualities for xylose fermentation will also be addressed. With the view that the ultimate use of a recombinant xylose-fermenting strain of *S. cerevisiae* is a large-scale industrial process, which requires stable and robust fermentative yeast, recent attempts to engineer metabolically polyploid strains with subsequent random mutagenesis will be summarized. The use of mathematical models to analyze the xylose metabolism will be discussed. First, the composition of the substrates for which these strains have been developed will be described.

2
Substrate Composition

The substrates used for xylose fermentation are lignocellulose hydrolysates, in particular hemicellulose hydrolysates. These contain a mixture of monosaccharides as well as various low molecular weight compounds which may inhibit both growth and ethanolic fermentation by recombinant *S. cerevisiae*. The final composition of lignocellulose hydrolysates depends on the raw material and how it has been physically, chemically, and biochemically treated to release the fermentable sugars. In this section, the composition of the lignocellulosic raw materials and the processes used to generate lignocellulose hydrolysates will be discussed in relation to the formation of fermentation inhibitors.

2.1
Composition of Hemicellulose Hydrolysates

Lignocellulosic materials consist of three major components: cellulose, lignin, and hemicellulose (see Fig. 2). Cellulose is a homopolysaccharide composed of D-glucose, or, more precisely, β-D-glucopyranose arranged as repeating units of cellobiose (for convenience the names of the open forms of the sugars will be used below). The cellulose has the form of insoluble fibers known as microcrystalline cellulose, interrupted by short amorphous regions. Lignin is a complex aromatic polymer synthesized from phenylpropanoid precursors. Hemicelluloses are branched heteropolysaccharides composed of hexoses, pentoses, and uronic acids.

The proportions of the monosaccharides obtained in hemicellulose hydrolysates will vary depending on the choice of raw material and the hydrolysis

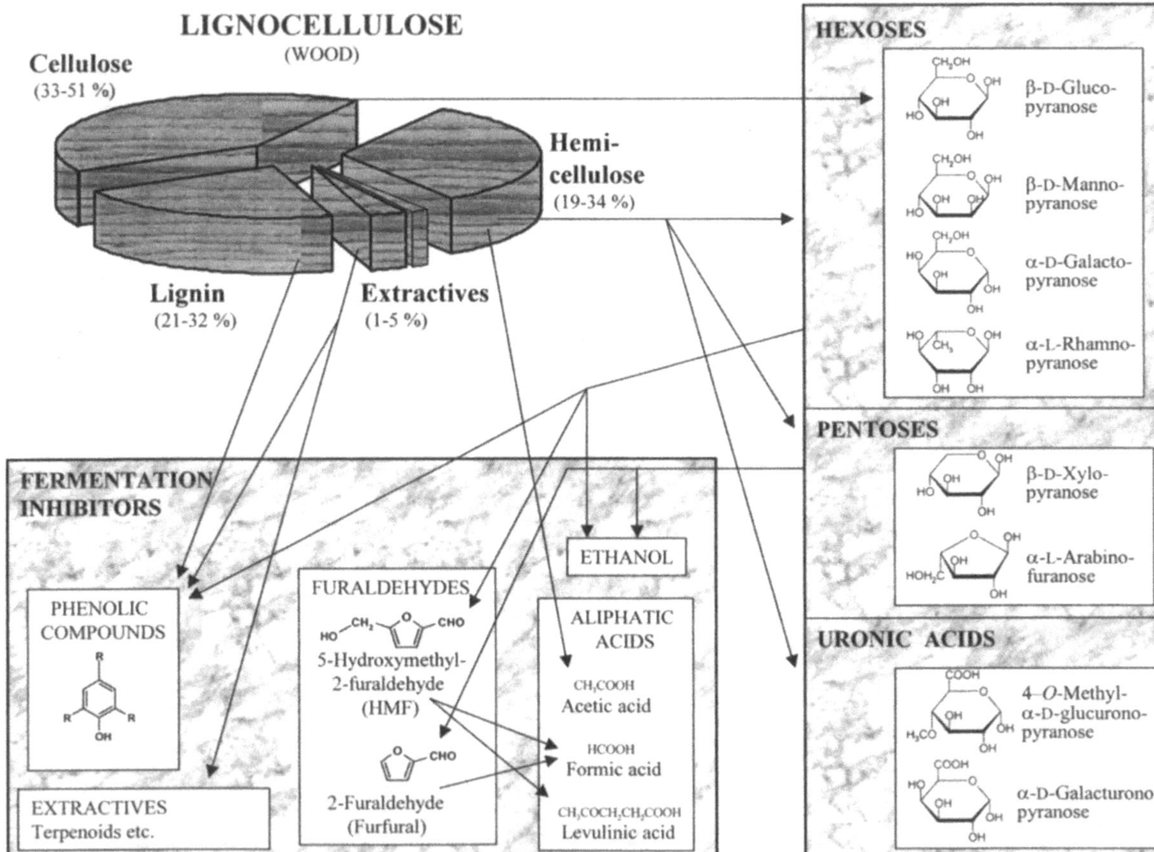

Fig. 2. Fermentation inhibitors in lignocellulose hydrolysates. Some of the possible pathways by which the inhibitors are generated, either by acid hydrolysis or in the ethanolic fermentation, are indicated in the figure

procedure. In general, hardwood hydrolysates contain a high proportion of xylose, while softwood hydrolysates are rich in mannose [57, 58].

In hardwood, the xylan, or glucuronoxylan, consists mainly of D-xylose residues, most of which are acetylated, and 4-*O*-methyl-D-glucuronic acid residues (Fig. 2). Minor amounts of other constituents, such as L-rhamnose and D-galacturonic acid, can also be found. The molar ratio of 4-*O*-methyl-α-D-glucuronic acid, acetyl groups, and xylose has been estimated to be 1:7:10 [59]. Hardwood also contains small amounts (2–5% dry weight (DW)) of glucomannan, composed of D-glucose and D-mannose.

In softwood, mannan is the dominating type of hemicellulose and can account for ~20% of the DW. The typical softwood mannan is a glucomannan with varying contents of D-galactose. A mannan with high galactose content is referred to as a galactogluco mannan. Softwood also contains arabinoglucuronoxylan, which is composed of L-arabinose, 4-*O*-methyl-α-D-glucuronic acid, and D-xylose in the molar ratio 1.3:2:10 [59].

Agricultural residues, such as wheat straw and sugarcane bagasse, contain large amounts of xylan (Table 1), some arabinan, and only very small amounts of mannan. Acid hydrolysis of wheat straw and sugarcane bagasse has been found to result in hydrolysates in which glucose and xylose together make up 95% or more of the recovered monosaccharides [57]. Corn fiber hemicellulose is basically an arabinoglucuronoxylan containing a very high proportion of pentose residues (xylose and arabinose) and, in addition, some hexoses and uronic acids, such as galactose and 4-*O*-methylglucuronic acid. The sugars obtained from corn fiber are mostly glucose (25–38%), xylose (30–41%), and arabinose (21–28%) [60].

2.2 Hydrolysis of Lignocellulose Polysaccharides

Monosaccharides are formed from the lignocellulose polysaccharides, hemicellulose and cellulose, by using either acid or enzymatic hydrolysis. Acid hydrolysis of lignocellulosic materials can be performed as a pretreatment stage, using sulfur dioxide or sulfuric acid [61, 62], resulting in a hemicellulose hydrolysate. In the second step, the solid residue is hydrolyzed using acid or cellulolytic enzymes in order to release glucose from cellulose. When cellulolytic enzymes are employed, the hydrolysis of the cellulose may be performed as a simultaneous saccharification and fermentation (SSF) process [63, 64].

Enzymatic saccharification steps usually involve enzymes obtained from an external source, such as cellulolytic enzymes from the filamentous fungus *Trichoderma reesei* [65]. The yeast *S. cerevisiae* is not known to hydrolyze efficiently either cellulose or hemicellulose. Much research has been conducted during the past two decades on the heterologous expression of genes encoding cellulolytic and hemicellulolytic enzymes in *S. cerevisiae*, working towards the potential consolidation of lignocellulosic hydrolysis and fermentation by a single microorganism, such as *S. cerevisiae*.

The interwoven nature of cellulose and hemicellulose in lignocellulosic raw material necessitates partial hydrolysis of cellulose to expose the hemicellulose fraction (mannans and xylans) to enzymatic hydrolysis. The enzymatic hydrolysis of crystalline cellulose requires the synergistic action of three different types of cellulolytic enzymes: endo-β-1,4-D-glucanases (endoglucanases), cellobiohydrolases, and cellobiases (β-D-glucosidases). Endoglucanase genes of both bacterial and fungal origin have been expressed in *S. cerevisiae*, generating recombinant yeasts that efficiently degrade glucans and amorphous cellulosic materials [66–71]. Cellobiohydrolase genes have been expressed with only limited success. Recombinant *S. cerevisiae* strains producing fungal cellobiohydrolases catalyzed only partial solubilization of microcrystalline cellulose [71–74]. However, various recombinant β-glucosidase *S. cerevisiae* strains have been constructed that allow the effective conversion of cellobiose and shorter oligosaccharides, released by endoglucanases and cellobiohydrolases from cellulose, to fermentable D-glucose [71, 74–79].

The enzymatic hydrolysis of galacto(gluco)mannans, present particularly in softwoods, is accomplished through the action of endo-β-1,4-mannanases which randomly cleave the β-mannosidic linkages within the main chain, together with β-1,4-mannosidases and α-1,6-galactosidases. Genes for mannan-degrading enzymes have been expressed and characterized in *S. cerevisiae* [80–83]; however, these genes have not yet been co-expressed in *S. cerevisiae* to develop a mannan-utilizing recombinant yeast. The effective enzymatic release of D-xylose from xylan requires the simultaneous production of several hemicellulases: β-1,4-xylanases (xylanases) and side-chain-splitting enzymes such as α-L-arabinofuranosidases, α-glucuronidases, and acetyl and phenolic esterases. The final hydrolysis of xylobiose and small *xylo*-oligosaccharides to D-xylose requires the action of β-xylosidases. Numerous xylanases and side-chain-splitting enzymes have been successfully produced in *S. cerevisiae* [84–93]. Recombinant *S. cerevisiae* producing β-xylanase II from *T. reesei* yielded both D-xylose and xylobiose as end products from birchwood xylan, with xylobiose as the major product [93]. However, *S. cerevisiae* producing both *T. reesei* β-xylanase II and *Aspergillus niger* β-xylosidase released substantially more D-xylose than xylobiose as end-products, representing a conversion of xylan to D-xylose of more than 40%. The introduction of the genes for these xylanolytic enzymes into xylose-utilizing recombinant *S. cerevisiae* strains [94] could pave the way for the bioconversion of the hemicellulose fraction of lignocellulosic materials to ethanol, as a microbial phenomenon by a single fermentative microorganism, *S. cerevisiae*.

2.3
Fermentation Inhibitors in Lignocellulose Hydrolysates

Only recently has the fermentative performance of recombinant xylose-utilizing *Saccharomyces* strains been investigated in lignocellulose hydrolysates [95–97]. In addition to dealing with varying proportions and concentrations of the monosaccharides, the metabolically engineered strains will be affected by a variety of low molecular weight compounds present in the hydrolysates. Individually, and in synergy, these may both stimulate and inhibit fermentation, which makes it difficult to predict the fermentability of a particular hydrolysate.

The fermentation inhibitors generated by acid hydrolysis of lignocellulose can roughly be divided into aromatic compounds (many of which are phenolics), furaldehydes, aliphatic acids, and extractives (Fig. 2). In addition, ethanol generated during the fermentation process may reach concentrations that will negatively affect the fermenting microorganism. Fermentation inhibitors have been the topic of several reviews [17, 98, 99] and are only briefly surveyed here.

Phenolic compounds are formed by the degradation of lignin. Additionally, some of the extractives in lignocellulose are of a phenolic nature [59]. A third source is sugar-derived phenolics, which can be formed under acidic conditions at elevated temperatures (Fig. 2). Specific removal of low molecular weight phenolics from a willow hemicellulose hydrolysate and a spruce hydrolysate using a phenoloxidase has directly demonstrated the inhibitory effect of phenolic compounds on *S. cerevisiae* [100, 101]. Enzymatic detoxification methods also open up the possibility of developing inhibitor-resistant strains of *S. cerevisiae* by means of genetic engineering.

Inhibitory furaldehydes in lignocellulose hydrolysates include 2-furaldehyde (furfural) and 5-hydroxymethyl-2-furaldehyde (HMF) (Fig. 2). The concentrations of furfural and HMF in lignocellulose hydrolysates are highly dependent on the raw material and on the conditions used for acid hydrolysis. Softwood acid hydrolysates contain low amounts of furfural compared with HMF [58]. Hardwood hydrolysates, which contain high concentrations of pentoses, the precursors to furfural, contain more similar amounts. Several recent investigations [102–105] deal with the effect of the furaldehydes on *S. cerevisiae* and the conversion of furfural to furfuryl alcohol and HMF to 5-hydroxymethyl-furfuryl alcohol by *S. cerevisiae*. The presence of the furaldehydes causes lag phases in the formation of biomass and ethanol.

Acetic acid is formed from acetyl groups in hemicellulose during acid hydrolysis, as well as during steam pretreatment without the addition of mineral acids. Formic and levulinic acids can be formed by the degradation of furans (Fig. 2). Low amounts of acetic acid increase the ethanol yield at the expense of the biomass yield [106, 107]. The concentration of the undissociated form of acetic acid should not exceed 5 g l^{-1} (0.08 mol l^{-1}) to permit growth of *S. cerevisiae* under anaerobic conditions [107]. Low concentrations (up to approx. 0.1 mol l^{-1}) of acetic, formic and levulinic acid were found to result in increased ethanol yield in oxygen-limited fermentation with *S. cerevisiae* [103], while higher amounts were inhibiting. Hardwood acid hydrolysates generally contain high amounts of acetic acid compared with softwood acid hydrolysates [58]. The resistance against fermentation inhibitors by different strains of *S. cerevisiae* is an important consideration in the selection of strains for metabolic engineering. This is further discussed in Sect. 6.

3 Xylose Transport

The first metabolic step in the fermentation of xylose is the uptake of the sugar through the plasma membrane. Although *S. cerevisiae* is able to transport xylose into the cell, it is not geared for efficient uptake of xylose at low concentrations or in the presence of glucose. The currently available data on xylose transport by *S. cerevisiae* and a comparison of natural xylose-fermenting yeasts with *S. cerevisiae* suggest that the xylose uptake in *S. cerevisiae* must be improved in order to construct an efficient xylose-fermenting strain.

3.1 Xylose Transport in *S. cerevisiae*

Xylose is taken up in *S. cerevisiae* by the glucose transporters [108]. These are permeases that transport sugars by facilitated diffusion [109] (Fig. 3), and have about two orders of magnitude lower affinities towards xylose than glucose (Table 2), which leads to competition between glucose and xylose when simultaneously present in the fermentation medium. When these two sugars were co-fermented by recombinant *S. cerevisiae* the uptake of xylose (15 g l^{-1}) was severely retarded until the glucose concentration fell below 10 g l^{-1} [110].

The xylose uptake may have a large impact on the xylose fermentation rate, even in the absence of glucose. Sugar transport is also one of the main rate-controlling steps in glucose fermentation by *S. cerevisiae* [111–113]. More recently, metabolic control analysis (MCA, see Sect. 9) of the closely related *S. bayanus* showed that the uptake has 60–100% control over the glycolytic flux in cells harvested at the diauxic shift [114]. Since *S. cerevisiae* takes up xylose with low affinity, the transport step should pose a limitation on the flux, at least at low substrate concentrations. In contrast, zero *trans*-influx of xylose was observed to be 30 times higher than the actual xylose consumption rate at the same concentration (100 mmol l^{-1}) in a recombinant XR- and XDH-expressing strain [46]. Heterologous expression of a tobacco monosaccharide-proton symporter

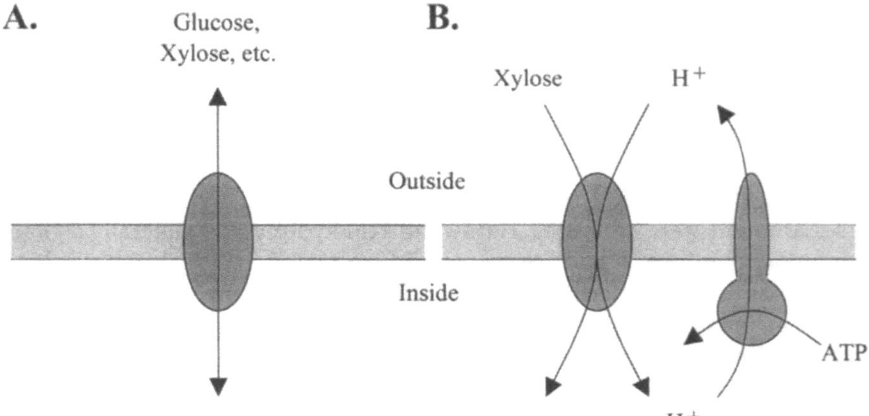

Fig. 3 A, B. The two mechanisms of xylose uptake by yeast: **A** facilitated diffusion – the driving force is the concentration gradient between the medium and the cytosol – these transporters generally have a broad substrate range; **B** proton-xylose symport. – the driving force is the proton motive force, which is maintained by the plasma membrane proton-ATPase. Adapted from [109]

Table 2. Comparison of the yeast xylose transporters

Organism	Low affinity system K_M (mmol l^{-1})		High affinity xylose uptake K_M (mmol l^{-1})	Reference
	Xylose	Glucose		
C. shehateae	125	2	1	[119]
C. utilis	67.6	ND	1.9	[120]
D. hansenii	140	18.5–25.0	0.8	[121]
P. heedii	40–50	ND	0.1	[122]
P. stipitis	380	ND	0.9	
P. stipitis	2–3[b]	0.2–0.7[b]	0.04–0.07	[123]
P. stipitis	19–80	1.9–14	0.2–3.2	[124]
R. glutinis	18	ND	0.56	[125]
S. cerevisiae	160	ND	–[a]	[108]
	1460	ND		
S. cerevisiae	190	1.5	–[a]	[46]
	1500	35		

ND: Not determined.
[a] Not present.
[b] May be high affinity glucose transporter. See text for details.

with good affinity towards xylose [115] had no effect on the xylose fermentation rate in a similar recombinant strain [116]. However, these strains produced low XDH activities [117] and XK was not overproduced, thus the xylose-metabolizing pathway had severe limitations (see Sects. 4 and 5). In this context, it is worth noting that XR, the first enzyme in the xylose-utilizing pathway, has a low affinity towards xylose (K_M is 68 mmol l^{-1} or 97 mmol l^{-1}, depending on the cofactor [118]), which means that high intracellular concentrations of xylose are necessary for efficient utilization. Calculations based on the zero *trans*-influx measurements do not account for the significant efflux of xylose under physiological conditions due to facilitated diffusion working in both directions. For glucose-derepressed cells the net glucose influx is only half of the uptake rate determined in vitro, because of the build-up of intracellular glucose [113].

3.2 Xylose Transport in Natural Xylose-Utilizing Yeasts

Most of the natural xylose-utilizing yeasts have at least two kinetically distinct xylose transport systems (Table 2) [119–125]. The low-affinity transporter is generally shared with the structural analog glucose, while the high-affinity transporter is specific for xylose. High-affinity systems symport xylose together with a proton, using the proton motive force (Fig. 3). The low-affinity systems, on the other hand, are generally thought to transport xylose by a facilitated diffusion process driven by the concentration gradient [119, 121]. A low-affinity proton symporter reported from *P. stipitis* displayed a K_M of 2–3 mmol l^{-1} [123], whereas another investigation [122] reported a markedly different low-affinity system (K_M = 380 mmol l^{-1}). Since the low-affinity xylose-proton symporter was strongly inhibited by glucose and starvation changed its kinetic constants similarly to the glucose-proton symporter, it may be an unspecific glucose-proton symporter. In *Candida utilis*, the low-affinity xylose transport was inhibited by D_2O, protonophores, and ATPase inhibitors, suggesting that it might be a proton symport [120]. On the other hand, proton movement was not observed and diethylstilbestrol, a potent ATPase inhibitor, had no effect on the xylose transport rate.

Recently, three genes from *P. stipitis* coding for low-affinity glucose transporters have been cloned and sequenced [124]. The glucose-induced *SUT1* has a K_M = 1.5 mmol l^{-1} for glucose and K_M = 149 mmol l^{-1} for xylose. It seems to be the major contributor to the low-affinity component of the glucose and xylose transport, which is evident from the lack of a low-affinity component in the *sut1* disruption strain grown on glucose. *SUT2* and *SUT3* have somewhat higher affinities for xylose (K_M = 49 mmol l^{-1} and 103 mmol l^{-1}, respectively), but they are only expressed under fully aerobic conditions, and have a substrate-concentration-modulated affinity for glucose. Such phenomena have previously been observed for the *S. cerevisiae HXT2* [126].

It is noteworthy that the Michaelis constant for the low-affinity xylose transport in *P. stipitis* is different from that for the *SUT1* determined in an *hxt1–7* deletion strain of *S. cerevisiae* [124]. This suggests that the apparent low affinity component is in fact the superposition of several individual transporters. In *S. cerevisiae* three or more hexose transporter genes are transcribed at the same time [127], yet only two components of the glucose uptake system can be kinetically distinguished [128]. It is, therefore, expected that the kinetic constants for xylose transporters from various species will be refined, once the corresponding genes are cloned and expressed in a model system such as the *hxt1–7 S. cerevisiae* strain.

3.3
Engineering Xylose Transport in *S. cerevisiae*

The low affinity of *S. cerevisiae* transporters for xylose and the inhibition by glucose underlines the necessity of engineering this metabolic step. However, there are no known nucleotide sequences of yeast or fungal origin coding for a xylose transporter with suitable properties. Cloning and characterization of yeast xylose transporters may be greatly facilitated by the use of hexose transporter deleted *S. cerevisiae* strains. The cloning of the low-affinity glucose/xylose transporters from *P. stipitis* was accomplished by functional complementation [124] of the glucose uptake by an *hxt1–7* deletion strain [129]. A similar approach was used for cloning of an unspecific monosaccharide transporter from the filamentous fungus *T. reesei* [130]. Recently, an *S. cerevisiae* strain, EBY.VW4000, has been developed with all hexose transporters deleted (*hxt1–17, gal2, stl1, agt1, ydl247w, yjr160c*) [131]. This strain is expected to become a highly useful tool for cloning of specific high-affinity xylose transporter genes from natural xylose-utilizing organisms.

4
The Conversion of Xylose to Xylulose

In yeast and filamentous fungi, xylose is converted to xylulose in two steps, where the first reaction is catalyzed by xylose reductase (XR) and the second by xylitol dehydrogenase, (XDH) (Fig. 1) [24]. Procaryotic organisms use a xylose isomerase (XI) to perform the conversion in one step [132].

4.1
Xylose Reductase (XR)/Xylitol Dehydrogenase (XDH)

XRs from different microorganisms have been characterized and they share a common feature in their preference for NADPH as a cofactor. The unspecific aldose reductase from *S. cerevisiae* having XR activity [37] and XR from *C. utilis* [133] exclusively use NADPH, whereas XR from *P. stipitis* [118, 134] and *Candida tenius* [135] are also able to use NADH. The ratio of the specific activity of XR from *P. stipitis* using NADH and NADPH separately was around 0.65, regardless of the oxygen tension in the medium [33]. *P. tannophilus* produces two isoenzymes of XR of which one can use both NADH and NADPH and the other is strictly NADPH dependent [136]. The expression of the different isoforms is dependent on the oxygenation level such that a low level of oxygenation favors the enzyme using both cofactors [137]. The equilibrium constant for the reduction of xylose to xylitol has been estimated to be 0.575×10^3 (M^{-1}) at pH 7 [118], thus favoring xylitol formation.

Unlike XR, XDH from all microorganisms studied almost exclusively uses NAD^+ as a cofactor [39, 133, 138, 139]. The equilibrium constant at pH 7 is 6.9×10^{-4} (M); thus this reaction also favors xylitol formation [140].

S. cerevisiae has been transformed with the *P. stipitis* genes *XYL1* and *XYL2* coding for XR and XDH, respectively [46, 47, 141]. The choice of *P. stipitis* as the donor organism was based on its capability to utilize NADH in the xylose reduction step. Attempts to ferment xylose to ethanol with these recombinant *S. cerevisiae* producing XR/XDH have resulted in low ethanol yield and considerable xylitol by-product formation. This has been ascribed to the unfavorable thermodynamic properties of the reactions [140] and the fact that the first reaction preferably consumes NADPH, whereas the second reaction exclusively produces NADH. When less NADH is consumed in the XR reaction, then less NAD^+ is available for the XDH reaction. If the amount of NAD^+ is insufficient, xylitol is produced and excreted [133].

In the following sections, measures to circumvent the cofactor imbalance generated in the first two steps of xylose metabolism in recombinant *S. cerevisiae* expressing XR and XDH will be discussed.

4.1.1
Activity Ratios for XR and XDH

To compensate for the unfavorable equilibrium constants, yeast strains with higher XDH activity than XR were constructed [142]. Product formation was studied in strains with ratios of XR:XDH enzyme activities ranging from 17.5 to 0.06. The strains were cultivated in shake flasks with minimal medium under oxygen-limited conditions. The strain with the highest XR:XDH ratio produced xylitol with a yield of 0.82 g xylitol g $xylose^{-1}$, whereas no xylitol was formed by the strain with the lowest ratio.

In a theoretical approach to optimizing the levels of XR and XDH and also XK, the enzyme phosphorylating xylulose to xylulose 5-phosphate, a kinetic model including the three enzymes was constructed [143]. Based on reported kinetic data for the three enzymes, the optimal XR:XDH:XK ratio was determined to be 1:10:4 for minimal xylitol formation. Experiments confirmed that a decreasing XR:XDH ratio decreased xylitol and acetate formation, whereas the formation of ethanol increased. Overproduction of XK enhanced the specific xylose consumption [143].

4.1.2
Protein Engineering – Fusion Protein

Xylitol formation would decrease if recycling of NADH/NAD^+ could take place in a single enzyme where NADH was oxidized at the XR site and reduced at the XDH site. Xylitol would then remain an enzyme-bound intermediate and the high microenvironmental concentration of NADH around the XR site would favor the utilization of NADH. A series of XR and XDH fusion proteins were constructed [144]. The specific activities of XR and XDH depended on the order in which the two polypeptides were coupled in the hybrid protein, as well as on the length and composition of the connecting peptide. To obtain both XR and XDH activity, XDH had to be at the N-terminus and XR at the C-terminus of the fusion protein. Constructs with the opposite order lacked XR activity. The specific XDH activity increased threefold in the construct containing a linker consisting of 12 amino-acid residues, compared with a 7-residue linker, while the XR activity remained constant.

The fusion protein exhibited only one tenth of the XR and XDH activity exhibited by the two enzymes when expressed from separate genes. When the fusion protein was co-expressed with the individual XR and XDH enzymes, aggregates composed of the fusion protein and the separate XR and XDH subunits were confirmed by gel chromatography. This construct had specific XR and XDH activities similar to the individually expressed enzymes. Recombinant *S. cerevisiae* strains harboring the fusion protein aggregate utilized xylose under oxygen-limited conditions in a defined medium and produced less xylitol than a strain expressing the enzymes separately.

4.1.3
Protein Engineering – Site-Specific Mutagenesis

Protein engineering has also been used to alter the co-factor preferences of XR and XDH. Inhibition studies of *P. stipitis* XR suggested that histidine and cysteine residues might be involved in co-factor binding [145]. Using site-directed mutagenesis, the three cysteine residues were individually changed into serine residues [146]. The three mutant forms of XR showed activity when expressed in *E. coli*, but only at levels 50–70% lower than that of the wild type. The affinities for xylose, NADPH and NADH did not vary significantly and it was concluded that none of the cysteine residues directly participates in the binding of co-factor.

Yeast xylose reductases as well as mammalian aldo-keto reductases contain a strictly conserved binding motif for NADPH (Ile-Pro-Lys-Ser). It has been suggested that the 2′-phosphate group of NADPH binds to the lysine residue in human aldose reductase. When this group was changed into a methionine residue, using site-specific mutagenesis, the resulting enzyme lost 80–90% of its specific activity and the affinity for xylose decreased by more than tenfold [147]. The affinity for NADPH decreased, but remained constant for NADH. There are, to date, no reports of the expression of any of the mutated forms of XR in *S. cerevisiae.*

Attempts have also been made to alter the co-factor specificity of XDH towards $NADP^+$ instead of NAD^+. Through sequence analysis of XDH, a coenzyme-binding domain, conserved in most examined NAD^+-dependent dehydrogenases was localized [148]. In the coenzyme-binding domain of horse liver alcohol dehydrogenase, an aspartate residue and a lysine/arginine residue are responsible for the interaction with the adenine ribose of NAD^+. The steric properties of the aspartate residue and the repulsion between the negatively charged groups of the phosphate of $NADP^+$ and the carboxyl group of aspartate prevent binding of $NADP^+$. The specificity for NAD^+ decreased when the aspartate residue was changed to a glycine residue. Although steric and electrostatic hindrances were avoided through this substitution, the affinity for $NADP^+$ remained unchanged. Furthermore, the specific activity of the mutated XDH decreased to half of that of the original enzyme.

The putative binding motif of an $NADP^+$-dependent alcohol dehydrogenase of *Thermoanaerobium brockii* was introduced into XDH from *P. stipitis* [148]. The resulting enzyme showed a specific activity of 31% of that of the unaltered enzyme and, as above, the affinity for NAD^+ decreased ninefold whereas the affinity for $NADP^+$ remained unchanged. When the altered enzyme was expressed together with a xylose reductase in *S. cerevisiae*, growth was observed on xylose minimal medium plates.

4.1.4
Xylitol Transport

The equilibrium constant for the conversion of xylitol to xylulose favors xylitol formation. The equilibrium would shift towards xylulose formation if the intracellular concentration of xylitol were increased. This could be achieved by limiting the xylitol excretion. The *S. cerevisiae* gene *FPS1* encodes a channel protein, Fps1p, responsible for the facilitated diffusion of glycerol [149]. Its main role is to control the cellular osmoregulation by the accumulation and release of glycerol [150]. However, it has recently been demonstrated that xylitol inhibited the glycerol transport by this protein, suggesting that Fps1p also transports xylitol [151]. When the *FPS1* gene was deleted in an *S. cerevisiae* strain harboring the *P. stipitis* genes for XR and XDH, the excretion of xylitol decreased and ethanol production increased compared to the parental strain (B. Hahn-Hägerdal, unpublished work). Furthermore, it was confirmed that the intracellular xylitol concentration increased.

4.1.5
Oxygen Utilization

When present, oxygen is used as an electron acceptor in the electron transport chain regenerating NAD^+ for xylitol oxidation. In *P. stipitis*, the yield of ethanol from xylose increased with decreasing oxygen flux [33]. Recently, the impact of oxygen on xylose utilization by a recombinant *S. cerevisiae* strain harboring the *P. stipitis* genes for XR and XDH, as well as an overexpressed XK gene, was investigated in a series of continuous cultivation experiments on mixtures of 15 g l^{-1} xylose and 5 g l^{-1} glucose [143]. With increasing oxygenation, the ethanol yield, calculated, as grams of ethanol per gram of total carbohydrate uptake, remained approximately constant at around 0.34, whereas the yields of glycerol and xylitol decreased and more carbon was used instead for biomass synthesis.

4.2
Xylose Isomerase

The co-factor imbalance generated by the first two steps in xylose metabolism could be entirely circumvented if the conversion of xylose to xylulose were to be catalyzed by the prokaryotic enzyme xylose isomerase (XI, Fig. 1). D-Xylose (glucose) isomerase EC 5.3.1.5 catalyses the reversible isomerization of D-xylose and D-glucose to D-xylulose and D-fructose, respectively. XI does not require redox cofactors and cannot generate cofactor imbalance during anaerobic xylose utilization.

The different bacterial XIs fall into two distinct groups (Fig. 4) based on their physical properties and their sequence homology (reviewed in [152]). Enzymes from the high G+C Gram-positive bacteria (*Actinoplanes*, *Streptomyces*, *Arthrobacter* species) and *Thermus* species belong to group I. These enzymes have a molecular mass of approximately 45 kDa and exhibit alkaline pH optima. Group II enzymes comprise all the other XIs (for example *E. coli*, *Bacillus*, *Clostridium*, and *Thermotoga* species), including the only characterized eukaryotic XI from *Hordeum vulgare* [153]. The group II XIs have an extended *N*-terminal region and a molecular mass of approximately 50 kDa. Their pH optima are close to neutral.

4.2.1
Expression of Xylose Isomerase in S. cerevisiae

Early attempts to produce XI in *S. cerevisiae* have failed. Transformation with *Actinoplanes missouriensis* [53] and *Clostridium thermosulfurogenes* [55] *xylA* did not result in the expression of XI, although the specific mRNA was present. The heterologous expression of the *E. coli* [51, 52] and the *Bacillus subtilis* [53]

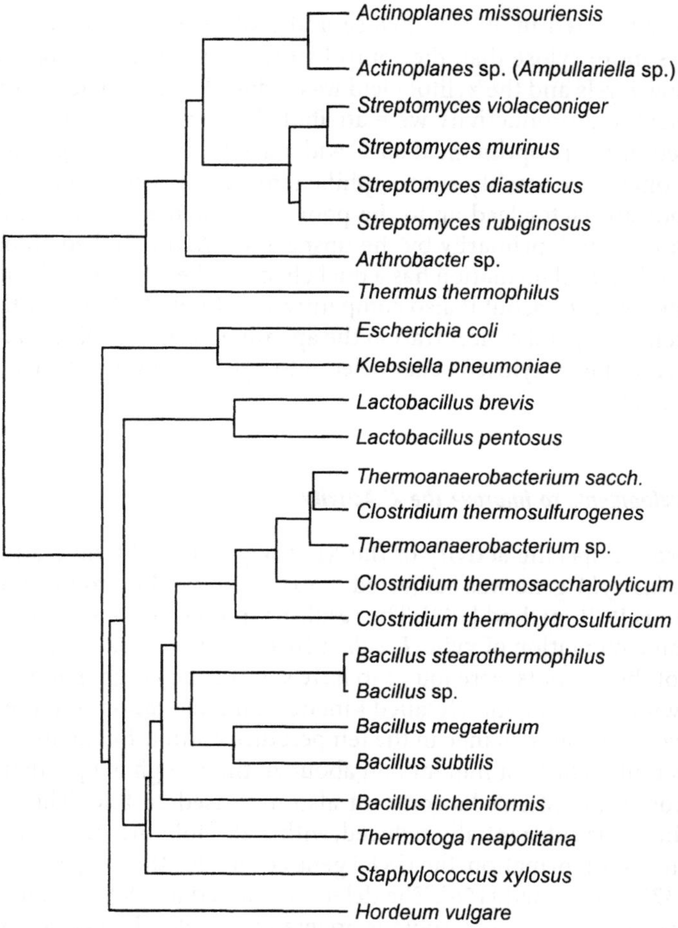

Fig. 4. Xylose isomerase dendrogram. Adapted from [56]

genes led to large amounts of mostly insoluble protein, which was catalytically inactive. It was speculated that improper protein folding, post-translational modifications, inter- and intramolecular disulfide bridge formation, or the yeast's internal pH caused the lack of activity [52]; however, none of these suggestions was verified. Post-translational modifications were experimentally excluded [52]. Disulfide bridge formation is rather unlikely, since the cytosolic environment of the yeast is known to be reducing [154]. Incompatibility of XI with the yeast's internal pH is clearly not the case, because the cytosolic pH is close to neutral [155]. In vitro refolding of the heterologously expressed *B. subtilis* XI resulted in a soluble protein with tertiary structure similar to the native one, but remained inactive [53].

XI from the thermophilic bacterium *T. thermophilus* was expressed successfully in *S. cerevisiae* [56]. SDS-PAGE and enzyme assay on cell extracts confirmed the presence and the activity of the enzyme. When the recombinant strain was cultivated in 30 g l^{-1} xylose under oxygen limitation, it consumed three times more xylose than the control strain. Ethanol and acetate were produced at low levels and the xylitol yield was reduced by half. The relatively poor ethanol yield and productivity were attributed to two factors. *T. thermophilus* XI has a temperature optimum at 85 °C with an activity of 1 U mg^{-1}, and the enzyme has only trace activity at mesophilic temperatures, 0.04 U mg^{-1} [56]. The other important factor leading to the poor performance of the strain was the formation of xylitol, primarily by the unspecific NADP*H*-linked aldose reductase [37, 156]. Xylitol formation has a dual effect on the ethanol yield; it not only leads to loss of carbon, but it also competitively inhibits XI [157]. With increasing intracellular xylitol concentration the apparent affinity of XI towards xylose decreases, and more xylose is channeled into xylitol, until the NADPH pool of the cell is depleted.

4.2.2
Recent Developments to Improve the XI Activity

To increase the specific activity of the XI at physiological temperatures, the *T. thermophilus xylA* gene was subjected to extensive random mutagenesis using an erroneous PCR method [158]. The resulting mutants were screened for functional complementation of *xylA⁻ E. coli* at 30 °C. Enzyme assays were performed and four of the mutants were found to have significantly higher activity at 30 °C than the wild-type enzyme. Detailed kinetic characterization of these four mutants showed no major change in the temperature optima, but an increase in the specific activity. The best mutant had about 70 times higher V_{max} than the wild type, although the Michaelis constant also increased 26-fold. The affinity towards xylitol was substantially reduced, with a 255-fold increase in K_i [158].

To limit xylitol formation, the *GRE3* gene coding for the unspecific aldose reductase [37] was deleted [156]. The deletion resulted in a 50% reduction of the xylitol formation in the anaerobic fermentation of 50 g l^{-1} xylose and 20 g l^{-1} glucose. The xylose uptake increased sixfold and the ethanol yield also showed a marked increase. Xylitol was probably formed from xylulose by an endogenous, unspecific, sugar alcohol dehydrogenase with XDH activity [39].

5
Hexose and Pentose Fermentation

Pentose sugars enter metabolism through the pentose phosphate pathway (PPP), where it has been suggested that xylulokinase [48] (Fig. 5) and transaldolase [49, 50] limit the flux of carbon to glycolysis in *S. cerevisiae*.

When the homologous gene for XK was overexpressed, seemingly contradictory results were obtained. Ethanol formation from xylose-glucose mixtures [94, 159, 160] and from xylulose increased [161], whereas growth on xylulose [162] and xylose consumption decreased [97]. Different host strains, different media (complex and defined), different aeration conditions (anaerobic, oxygen-limited, aerobic), and the presence/absence of hexose co-substrates may contribute to the apparent disagreement of the results. In addition, control of the

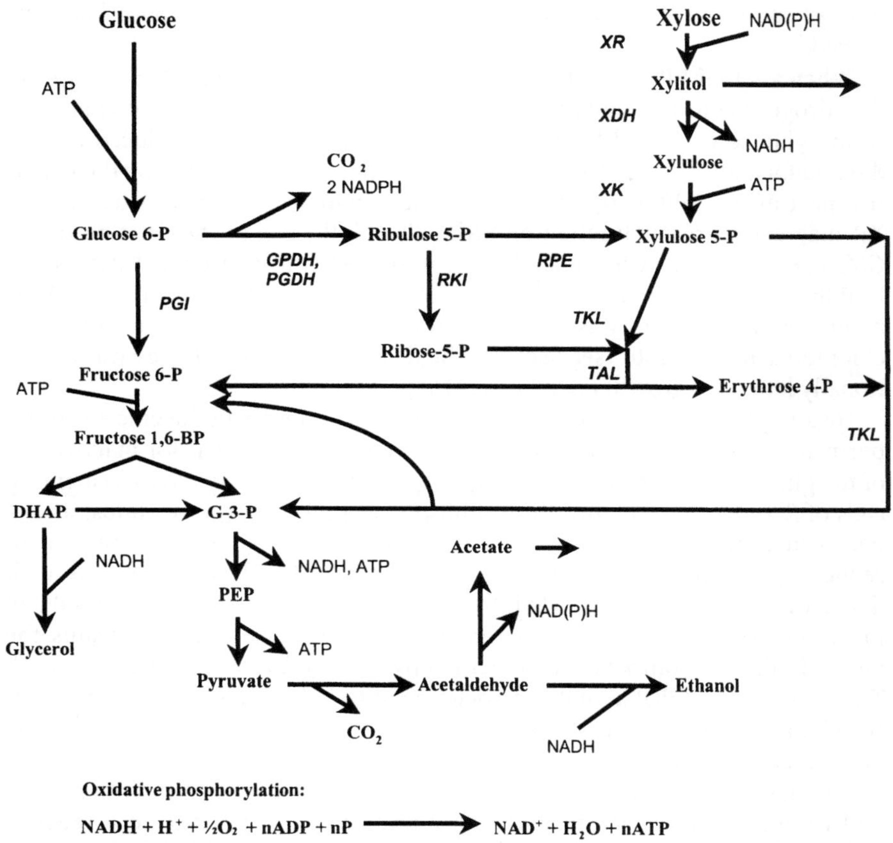

Fig. 5. A simplified metabolic scheme of ethanol formation from glucose and xylose. Enzyme abbreviations: *GPDH*: Glucose 6-phosphate 1-dehydrogenase, *PGDH*: Phosphogluconate dehydrogenase, *PGI*: Glucose 6-phosphate-isomerase, *RKI*: Ribose 5-phosphate isomerase, *RPE*: Ribulose phosphate 3-epimerase, *TAL*: Transaldolase, *TKL*: Transketolase, *XDH*: Xylitol dehydrogenase, *XK*: Xylulokinase, *XR*: Xylose reductase

expression level of XK may be crucial. This kinase consumes ATP at the beginning of a metabolic pathway. Metabolic modeling suggested that high, uncontrolled activity of such an enzyme leads to "substrate-programmed death" when the cell is depleted of ATP at a faster rate than ATP is regenerated [163]. In fact, the xylose consumption and ethanol formation rates were higher in a strain where XK was chromosomally integrated [160] than in a strain where XK was expressed from a multicopy plasmid [97]. In the chromosomally integrated strain, the XK activity was approximately 2 U mg^{-1}, and in the plasmid-carrying strain it was approximately 30 U mg^{-1}.

Overexpression of the endogenous *TAL1* gene encoding TAL enhanced aerobic growth of a recombinant strain of *S. cerevisiae* expressing *XYL1* and *XYL2* from *P. stipitis* [141]. Overexpression of *TKL1* encoding transketolase did not influence growth, but overexpression of both *TAL1* and *TKL1* improved aerobic growth even more. The overexpression of *TAL1* and *TAL1/TKL1* did not influence ethanol formation under the experimental conditions employed.

When another PPP gene, *GND1* encoding the enzyme gluconate 6-phosphate dehydrogenase in the oxidative part of the PPP, was deleted, the ethanol yield from xylulose increased by 30% [161]. This was ascribed to reduced carbon dioxide formation. NADPH for the XR reaction is provided either by the oxidative part of the PPP or by acetate formation from acetaldehyde. Alternatively, XR may use a greater fraction of NADH. The influence of the deletion of the *GND1* gene on the xylose metabolism is presently being investigated. Also in the PPP, the flow of carbon in the reaction catalyzed by the enzyme ribulose 5-phosphate epimerase (RPE) was shown to be very low when analyzed using a stoichiometric model [160] (see Sect. 9). The deletion of the *RPE1* gene was conditionally lethal for growth on xylulose [161].

Hexose phosphates are required for induction of the ethanologenic enzymes, pyruvate decarboxylase, and alcohol dehydrogenase, as well as for inactivation of the gluconeogenic fructose 1,6-bisphosphatase [164]. In xylulose-fermenting cells of *S. cerevisiae* fructose 1,6-bisphosphate (FBP) levels were almost an order of magnitude lower than in glucose-fermenting cells [44]. In strains with reduced phosphoglucose isomerase (PGI) activity [165] and in strains with deleted trehalose synthesis [166], fructose 6-phosphate and FBP accumulated intracellularly compared with parental strains. In these mutated strains the yield of ethanol from xylulose increased by 15% and 20%, respectively [161]. Thus, reduction of PGI activity and deletion of trehalose synthesis enhanced intracellular concentrations of FBP in xylulose-metabolizing cells to levels supporting ethanologenesis.

The inability of pentose sugars to support anaerobic growth in both natural and recombinant xylose-metabolizing yeasts has been ascribed to a reduced yield of ATP from pentose metabolism [18]. However, the yield of ATP per mole of carbon is the same for pentose and hexose metabolism (Fig. 5). Therefore it is rather the rate of pentose utilization that limits the rate of ATP generation during anaerobic metabolism. Under anaerobic conditions, the xylose flux was 2.2 times lower than the glucose flux in recombinant xylose-utilizing *S. cerevisiae* [94]. *P. stipitis* has been metabolically engineered for anaerobic glucose growth by expression of the *S. cerevisiae URA1* gene encoding dihydroorotate dehydrogenase, which catalyzes the conversion of dihydroorotate to orotate in the pyrimidine biosynthesis pathway [167]. This recombinant *P. stipitis* strain did not grow anaerobically on xylose. In *P. stipitis* the anaerobic sugar consumption rate is approximately 0.1 g g DW cells^{-1} h^{-1} for both xylose [33] and glucose [168], indicating that factors other than the rate of ATP generation limit anaerobic growth on xylose

6 Choice of Host Strain

The ultimate aim of developing xylose-fermenting strains is to use them in large-scale ethanol production from non-detoxified, non-sterilized lignocellulose hydrolysates. Under these conditions, stringent aeration control will not be possible. The hosts for developing xylose-fermenting strains must tolerate inhibitors generated in the production of hydrolysates (see Sect. 2.3). Strains tolerant of low pH are desirable since pH is a means of controlling contamination under non-sterile conditions. Ethanol tolerance is also of importance, since the inhibitory effect of ethanol is enhanced by the presence of hydrolysate inhibitors at low pH. Based on these considerations, strains of *Saccharomyces* were selected as being the most suitable hosts for developing efficient xylose-fermenting strains. *S. cerevisiae* produces ethanol from glucose independent of the aeration conditions. This yeast has been selected over thousands of years for rapid ethanol (wine, beer, distiller's yeast) and carbon dioxide (baker's yeast) production in high-osmolarity substrates containing acids and other fermentation inhibitory substances. This yeast has also been selected for its high ethanol tolerance [169–171]. Two independent studies comparing different yeasts [34] and comparing yeast with bacteria [172] have confirmed that *S. cerevisiae* outperforms all other organisms when their fermentative capacity is compared in non-detoxified lignocellulose hydrolysates.

S. cerevisiae strains with enhanced tolerance to spent sulfite liquor (SSL) were isolated from a pulp mill [173]. In these strains, the co-metabolism of glucose and galactose was enhanced by acetic acid at low pH. Despite considerable effort, it was not possible to identify any efficient xylose-utilizing strains in the pulp mill.

The other important factor in the selection of a potential host strain is the presence of an active and efficient PPP linking the introduced xylose-to-xylulose pathway to glycolysis. The fermentation of xylose-xylulose mixtures [43, 174] and xylulose [35, 161] has been compared. *S. cerevisiae* ATCC 24860, which is probably the same strain as CBS 8066, is by far the most efficient xylulose fermenter [35,43]. This strain has recently been transformed with the genes for the initial xylose-metabolizing enzymes and its xylose-fermenting capacity is presently being evaluated. However, strains isolated for their inhibitor tolerance such as *S. cerevisiae* isolated from SSL [34, 174] and for their acid tolerance, such as *Zygosaccharomyces bailii* and *Z. rouxii*, fermented xylulose poorly to ethanol [161]. Ongoing investigations will reveal the relative importance of the inhibitor tolerance and the efficiency of the PPP for the construction of efficient xylose-fermenting strains of *S. cerevisiae*.

7 Metabolic Engineering of Polyploid Strains

As discussed in Sect. 6, the development of xylose-fermenting strains will probably require industrial isolates of *S. cerevisiae* as genetic hosts which exhibit high tolerance to the inhibiting environment of industrial substrates, and possibly a well developed PPP. However, such strains are prototrophic and not amenable to genetic manipulation using commonly applied auxotrophic markers. Furthermore, genetic breeding and the expression of heterologous genes in industrial prototrophic yeast strains are not only restricted by a lack of a genetic transformation systems, but industrial yeast strains are usually diploid or aneuploid and often sporulate poorly. An ideal gene transfer system for industrial yeast strains thus requires the absence of bacterial plasmid nucleotide sequences in the transformant, stable inheritance of the transformed gene, a high transformation efficiency, a wide application to taxonomically diverse industrial yeast strains, as well as a dominant resistant-selectable marker [175]. Dominant selective markers, such as resistance against a drug, have commonly been applied to industrial strains for the selection of transformants; for instance resistance to G418/geneticin [176], methylglyoxal [177], methotrexate [178], cycloheximid [179], copper [180, 181], chloramphenicol [182], killer toxin-production/resistance [183], and sulfometuron methyl (SM) [184].

The introduction of a heterologous xylose-utilizing pathway in *Saccharomyces* provides for a selection system applicable to industrial yeast strains. The transformation of prototrophic strains with genes encoding xylose-metabolizing enzymes would allow selection for growth on xylose as the sole carbon source and the subsequent isolation of xylose-utilizing transformants. An ethanol-tolerant *Saccharomyces* isolate engineered for xylose fermentation was *Saccharomyces* strain 1400, a hybrid of *S. diastaticus* and *S. uvarum*, which was selected for its enhanced temperature tolerance [185]. Multiple copies of the *XYL1* and *XYL2* genes from *P. stipitis* and the *XKS1* gene from *S. cerevisiae* were introduced into *Saccharomyces* strain 1400 under the control of glycolytic promoters. Initially, the three genes were expressed from 2μ-based, high-copy-number plasmids, but later from multiple copies integrated into the genome of *Saccharomyces* strain 1400. Transformants were obtained through selection for growth on xylose. The recombinant *Saccharomyces* strain 1400 expressing the xylose-utilizing genes from the 2μ-based plasmids was relatively stable for four to five generations in non-selective medium. However, the recombinant strain containing multiple copies of the xylose-utilizing genes in the genome proved to be genetically stable under non-selective conditions. The recombinant *Saccharomyces* strain 1400 was able to ferment xylose to ethanol in complex media [1, 159].

More recently, an ethanol-tolerant industrial yeast was used as host for the creation of a recombinant xylose-fermenting strain by expression of *XYL1*, *XYL2*, and *XKS1*, integrated into the yeast's genomic *HIS3* locus (Cordero Otero, unpublished results). The recombinant strain demonstrated ethanol production from xylose in a defined medium

8 Classical Breeding Techniques

So far, the use of recombinant DNA technology has been used for the construction of novel xylose fermenting strains of *S. cerevisiae*. To develop a robust industrial strain it may also be useful to combine this bottom-up approach with top-down approaches, such as protoplast fusion, generation of hybrids by mating, or random mutagenesis. Protoplast fusion has so far not been successful in the development of new traits in yeasts. When an ethanol tolerant *S. cerevisiae* strain was fused with auxotrophic xylose-fermenting strains of *P. stipitis* and *C. shehatae*, the fusants were able to use xylose, however, without the ethanol tolerance of *S. cerevisiae* [186]. Generating yeast hybrids, on the other hand, has been a useful strategy for combining industrially desirable traits, as has been demonstrated for the development of brewer's yeast and wine yeast [185]. This technique can be very useful in the future to combine different traits needed for ethanol production from lignocellulose, such as efficient xylose conversion to ethanol and tolerance to lignocellulose hydrolysate inhibitors.

Random mutagenesis, linked with powerful selection and isolation protocols, is a very powerful top-down tool to generate new strains with desired traits. Random mutagenesis in yeasts is often induced by treating cells with mutagens to increase the mutation frequency. The two most common mutagens used with yeast cells are alkylating agents (*N*-methyl-*N'*-nitro-*N*-nitrosoguanidine, MNNG; ethylmethane sulfonate, EMS) and ultraviolet light (UV). Alkylating agents are highly specific in their action producing exclusively transitions at G · C sites [187], while UV light is an efficient mutagen that produce a greater range of substitutions, particularly at T-T pairs, including transitions and transversions [188]. Choosing an optimal dose usually requires balancing the competing needs for a high mutation frequency, and a reasonably high rate of survival (between 10% and 50%). For the development of industrial strains with new traits, the use of multiple mutagenesis cycles to introduce multiple mutations may be advantageous.

A recombinant yeast strain expressing the *XYL1* and *XYL2* genes of *P. stipitis* from an episomal plasmid containing the G418 resistance marker (kan^R) were subjected to EMS mutagenesis (50% survival rate) and transferred 16 times to fresh xylose medium under G418 selection [189]. The cultures were monitored for enhanced cell growth on xylose as carbon source and two mutants were retained. One mutant, IM2, exhibited a growth rate three times higher than the parental strain. The ethanol yield and productivity increased 1.6- and 2.7-fold, respectively. Closer analysis revealed that multiple integration of the *XYL1* and *XYL2* genes into the yeast chromosome took place in mutant IM2 and that the mutant showed higher xylulokinase activity [189].

Recently, a xylose-utilizing recombinant industrial yeast strain was treated with EMS (20–50% survival rate) prior to selection for improved xylose fermentation. Mutants with significantly enhanced growth and CO_2 production in xylose medium were retained and are presently being evaluated in fermentation of non-detoxified softwood hydrolysates (Cordero Otero, unpublished results).

9 Metabolic Modeling

Mathematical models allow the calculation of intracellular fluxes, the degree of control exerted by individual enzymes in a metabolic pathway, and the range of metabolite concentrations permitted to make a pathway thermodynamically feasible. Mathematical models can thus help elucidating the impact of genetic changes on cell physiology and aid in the rational design of future genetic engineering strategies. The topic has been extensively reviewed elsewhere [190] and here only the application of mathematical models in studies of xylose utilization will be highlighted.

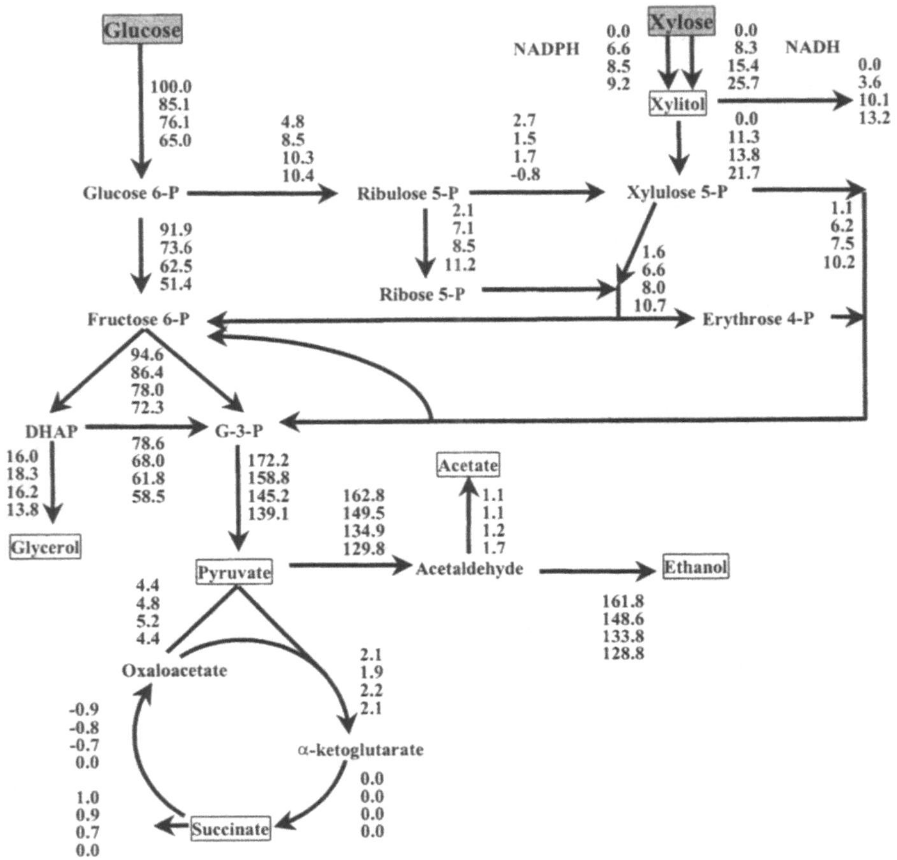

Fig. 6. Metabolic fluxes at a dilution rate of 0.06 h^{-1} and feed concentrations of xylose+glucose of: 0+20, 5+15, 10+10, and 15+5 g l^{-1}. All fluxes are normalized to a total specific sugar consumption of 100 mmol g^{-1} biomass h^{-1}. Grey boxes indicate substrate and substances enclosed by a box are excreted into the medium. From [160] with permission from John Wiley & Sons, Inc.

Intracellular fluxes are difficult to measure directly but can be calculated from stoichiometric data for the metabolic system together with measured extracellular fluxes. This is the basis of metabolic flux analysis (MFA) [191, 192]. It is then possible to see how split ratios at branch points and the flux to the product change under different environmental conditions. MFA has only recently been applied to investigate xylose metabolism. The intracellular fluxes in anaerobically grown recombinant *S. cerevisiae* TMB 3001 were calculated from chemostat data at different feed concentrations of xylose (Fig. 6) [160]. *S. cerevisiae* TMB 3001 is a recombinant CEN.PK strain that harbors the genes for XR, XDH, and an additional copy of the endogenous XK integrated into the genome [94].

The PPP flux increased with increasing xylose uptake and conversely less carbon was channeled to glycolysis. The model calculated that the ratio of

NADPH:NADH used in the first step of xylose utilization changed with environmental conditions. A larger fraction of xylose was reduced with NADH with increasing xylose uptake. The flux of xylitol channeled into the PPP corresponds well with the flux of xylose that is reduced with NADH in the XR reaction. The model thus verifies the hypothesis that xylitol excretion is due to a shortage of NAD^+, which is caused by the dual cofactor specificity in the XR reaction combined with the NAD^+ specificity of the XDH reaction [133]. The model suggests that compounds acting as electron acceptors such as furfural, present in non-detoxified lignocellulosic hydrolysate can regenerate NAD^+ for the XDH reaction and thus be beneficial for xylose utilization.

Intracellular fluxes can be calculated with higher accuracy by using ^{13}C-labeled substrate [193]. In this method, cells from chemostat cultures fed with ^{13}C-labeled substrate are hydrolyzed and the intracellular fluxes are then calculated from the labeling pattern of the amino acids together with knowledge of how they are derived from precursors in glycolysis, PPP, and the TCA-cycle. The method was used to analyze the fluxes in recombinant, xylose utilizing *Z. mobilis* [194]. From this MFA together with determinations of enzymatic activities, it was concluded that XK probably limits growth on xylose by this organism.

The results from MFA describe the intracellular fluxes, but cannot alone predict which enzymes exert most control of the flux through the pathway. In metabolic control analysis (MCA) [195, 196], a flux control coefficient (FCC) is calculated for each enzyme. The FCC varies between 0 and 1 and expresses the relative increase in flux through the pathway as a response to an infinitesimal change in enzyme activity. Hence, an enzyme with high FCC is a target for overexpression, since a small increase in its activity should result in a large flux increase. With this technique it was estimated that the transport of glucose to a large extent controls the flux through glycolysis in *S. cerevisiae* during anaerobic glucose fermentation [197]. However, MCA results must be treated with care, because small deviations in the experimental measurements will have a large impact on the FCCs [198]. So far, there are no reports on the use of MCA to study xylose utilization, but with the recent development of stable, recombinant xylose-utilizing strains of *S. cerevisiae* it will be of great interest and importance to use MCA to determine the distribution of control of the pathway from xylose to ethanol.

According to the second law of thermodynamics, spontaneous processes occur in the direction that increases the overall disorder (or entropy) of the universe. A more convenient criterion for a thermodynamically feasible reaction is a negative Gibbs free energy (ΔG). Gibbs free energy for a chemical reaction is influenced by the metabolite concentrations and this has been used to develop an algorithm that calculates a concentration range where all reactions in a pathway are feasible [199]. With this algorithm it was shown that the concentration of FBP must be high and the concentration of 1,3-bisphophoglycerate must be low to make glycolysis thermodynamically feasible. This concentration relation has to be fulfilled also in the conversion of xylose to ethanol (C. F. Wahlbom, unpublished results), and experimental analysis to confirm this is under way. As with MFA, no conclusions of how the pathway is controlled can be drawn. However, thermodynamic data together with intracellular metabolite concentrations at steady state can be used to calculate the FCCs [200]. The applicability was demonstrated in a study of the penicillin production pathway [200] and this technique could also be used to target genes for further genetic engineering of xylose utilizing *S. cerevisiae* to improve ethanol production from xylose.

10 Future Outlook

In a recent study the fermentative performance of recombinant *E. coli*, *Z. mobilis*, and *Saccharomyces* 1400 in pretreated corn fiber hydrolyzates was summarized [96]. The highest ethanol concentration, 34.7 g l^{-1}, was achieved with recombinant *E. coli* KO11, and the highest yield on consumed sugars and highest maximum volumetric productivity with *Saccharomyces* 1400, 0.50 g g^{-1} and 1.60 g l^{-1} h^{-1}, respectively. The figures are comparable to what is achieved in industrial hexose fermentation and would suggest that the development of recombinant xylose fermenting strains has come to a successful completion. However, these figures are hampered by the fact that corn fiber hydrolyzate contains hexose sugars, which strongly contribute to the yield on *consumed* sugars and the *maximum* volumetric productivity. Benchmarks for the development of recombinant xylose fermenting strains should include (i) yield on *total* sugars, (ii) *average* volumetric productivity (determined when all sugars are consumed or when the yield on total sugar is determined), and (iii) *specific* productivity (g ethanol g $cells^{-1}$ h^{-1}). The volumetric productivity relates to the design of the fermentation process and can be substantially improved by using high cell densities. The specific productivity is a benchmark for the fermentative performance of a particular strain. It is noteworthy that none of the recombinant xylose-fermenting strains have yet been demonstrated to work in industrial processes.

For xylose-fermenting recombinant strains of *S. cerevisiae* the yield is presently limited by the cofactor imbalance in the XR and XDH steps. This may be overcome by expressing mutants of XI where the activity is improved with respect to temperature optimum and xylitol inhibition. The rate of xylose fermentation could be improved by expressing high-affinity xylose transporters with a proton symport mechanism from naturally xylose-utilizing organisms. In addition, limitations of the PPP may be overcome by deleting or overexpressing genes for certain enzymatic steps. The inability of recombinant xylose-fermenting *S. cerevisiae* to grow anaerobically on xylose must also be addressed if these strains are to be applied in an industrial context.

So far, mainly bottom-up recombinant DNA technology has been used for the construction of novel xylose-fermenting strains of *S. cerevisiae*. This has been based on rational selection of genes to be manipulated and requires the target genes to be known. However, for some desired traits, such as anaerobic growth on xylose, the genes have not yet been identified and there is no simple assay to identify enzymes and proteins responsible for this phenotype. Then rational bottom-up recombinant DNA technology must be combined with classical top-down breeding techniques such as protoplast fusion, hybridization, and random mutagenesis linked to powerful selection and isolation protocols to generate mutants with desired phenotypes. The physiological characteristics of the mutants must be quantitatively assessed and the genes responsible for the altered phenotype can be identified by DNA micro array techniques [201] and proteome analysis [202]. These genes can then be rationally manipulated by bottom-up recombinant DNA technology to further improve the desired traits of selected phenotypes. This iterative strategy where bottom-up and top-down strain development techniques are combined with DNA micro arrays, proteomics, and quantitative physiology is expected to generate novel and efficient industrial xylose-fermenting strains of *S. cerevisiae*

Acknowledgements. The work to construct xylose-utilizing strains of *S. cerevisiae* at the Department of Applied Microbiology, Lund University, Sweden, and the Department of Microbiology, Stellenbosch University, South Africa, was financially supported by Energimyndigheten (Swedish National Energy Administration), STINT (The Swedish foundation for international cooperation in research and higher education), EU-contract BIO4-CT95-0107 ("Yeast Mixed Sugar Metabolism"), EU-contract QLK3-1999-00080 (BIO-HUG), and NRF (National Research Foundation), South Africa.

References

1. Ho NW, Chen Z, Brainard AP, Sedlak M (1999) Adv Biochem Eng Biotechnol 65:163
2. Aristidou A, Penttilä M (2000) Curr Opin Biotechnol 11:187
3. Rosenberg SL (1980) Enzyme Microb Technol 2:185
4. Tsao GT, Ladisch MR, Voloch M, Bienkowski P (1982) Process Biochem 17:34
5. Swedish National Energy Administration (1999) Energy in Sweden
6. Hall DO, Mynick HE, Williams RH (1991) Nature 353:11
7. Lynd LR, Cushman JH, Nichols RJ, Wyman CE (1991) Science 251:1318
8. Brown KS (1999) Science 283:1440
9. Hulme M, Barrow EB, Arnell NW, Harrison PA, Johns TC, Downing TE (1999) Nature 397:688
10. Zacchi G, Skoog K, Hahn-Hägerdal B (1988) Biotechnol Bioeng 32:460
11. Hinman ND, Wright JD, Hoagland W, Wyman CE (1989) Appl Biochem Biotechnol 20/21:391
12. von Sivers M, Zacchi G (1995) Biores Technol 51:43
13. Jeffries TW (1983) Adv Biochem Eng Biotechnol 27:1
14. Skoog K, Hahn-Hägerdal B (1988) Enzyme Microb Technol 10:66
15. Schneider H (1989) Crit Rev Biotechnol 9:1
16. Hahn-Hägerdal B, Hallborn J, Jeppsson H, Olsson L, Skoog K, Walfridsson M (1993) Pentose fermentation to alcohol. In: Saddler JN (ed) Bioconversion of forest and agricultural plant residues. CAB International, Wallingford, p 231
17. Olsson L, Hahn-Hägerdal B (1996) Enz Microb Technol 18:312
18. Jeffries TW, Shi NQ (1999) Adv Biochem Eng Biotechnol 65:117
19. Østergaard S, Olsson L, Nielsen J (2000) Microbiol Mol Biol Rev 64:34
20. Moniruzzaman M, Dien BS, Ferrer B, Hespell RB, Dale BE, Ingram LO, Bothast RJ (1996) Biotechnol Lett 18:985
21. Lawford HG (1988) Appl Biochem Biotechnol 17:203
22. Zhang M, Eddy C, Deanda K, Finkelstein M, Picataggio S (1995) Science 267:240
23. Deanda K, Zhang M, Eddy C, Picataggio S (1996) Appl Environ Microbiol 62:4465
24. Chiang C, Knight SG (1960) Nature 188:79
25. Chakravorty M, Veiga LA, Bacila M, Horecker BL (1962) J Biol Chem 237:1014
26. Smiley KL, Bolen PL (1982) Biotechnol Lett 4:607
27. Bicho PA, Runnals PL, Cunningham JD, Lee H (1988) Appl Environ Microbiol 54:50

28. Gong CS, Maun CM, Tsao GT (1981) Biotechnol Lett 3:77
29. Suihko ML, Enari TM (1981) Biotechnol Lett 3:723
30. Ueng PP, Gong CS (1982) Enzyme Microb Technol 4:169
31. Prior BA, Kilian SG, du Preez JC (1989) Process Biochem 24:21
32. Slininger PJ, Bolen PL, Kurtzman CP (1987) Enzyme Microb Technol 9:5
33. Skoog K, Hahn-Hägerdal H (1990) Appl Environ Microbiol 56:3389
34. Lindén T, Hahn-Hägerdal B (1989) Enzyme Microb Technol 11:583
35. Yu S, Jeppsson H, Hahn-Hägerdal B (1995) Appl Microbiol Biotechnol 44:314
36. Kolot FB (1980) Process Biochem 15:2
37. Kuhn A, van Zyl C, van Tonder A, Prior BA (1995) Appl Environ Microbiol 61:1580
38. Garay-Arroyo A, Covarrubias AA (1999) Yeast 15:879
39. Richard P, Toivari MH, Penttilä M (1999) FEBS Lett 457:135
40. Wang PY, Shopsis C, Schneider H (1980) Biochem Biophys Res Commun 94:248
41. Wang PY, Johnson BF, Schneider H (1980) Biotechnol Lett 2:273
42. Chiang LC, Hsiao HY, Ueng PP, Chen LF, Tsao GT (1982) Ethanol production from xylose by enzymic isomerization and yeast fermentation. In: Scott CD (ed) Biotechnol Bioeng Symp. Wiley, New York, p 263
43. Hahn-Hägerdal B, Berner S, Skoog K (1986) Appl Microbiol Biotechnol 24:287
44. Senac T, Hahn-Hägerdal B (1990) Appl Environ Microbiol 56:120
45. Hallborn J, Walfridsson M, Airaksinen U, Ojamo H, Hahn-Hägerdal B, Penttilä M, Keränen S (1991) Biotechnology (NY) 9:1090
46. Kötter P, Ciriacy M (1993) Appl Microbiol Biotechnol 38:776
47. Tantirungkij M, Nakashima N, Seki T, Yoshida T (1993) J Ferm Bioeng 75:83
48. Deng XX, Ho NW (1990) Appl Biochem Biotechnol 24/25:193
49. Ciriacy M, Porep H (1986) Conversion of pentoses to ethanol by baker's yeast. In: Magnien E (ed) Biomolecular engineering in the European community. Martinus Nijhoff, Dordrecht, The Netherlands, p 677
50. Senac T, Hahn-Hägerdal B (1991) Appl Environ Microbiol 57:1701
51. Ho NWY, Stevis P, Rosenfeld S, Huang JJ, Tsao GT (1983) Biotechnol Bioeng Symp 13:245
52. Sarthy AV, McConaughy BL, Lobo Z, Sundstrom JA, Furlong CE, Hall BD (1987) Appl Environ Microbiol 53:1996
53. Amore R, Wilhelm M, Hollenberg CP (1989) Appl Microbiol Biotechnol 30:351
54. Hallborn J (1995) PhD thesis, Lund University
55. Moes CJ, Pretorius IS, van Zyl WH (1996) Biotechnol Lett 18:269
56. Walfridsson M, Bao X, Anderlund M, Lilius G, Bülow L, Hahn-Hägerdal B (1996) Appl Environ Microbiol 62:4648
57. Puls J (1993) Substrate analysis of forest and agricultural wastes. In: Saddler JN (ed) Bioconversion of forest and agricultural plant residues. CAB International, Wallingford, p 13
58. Taherzadeh MJ, Eklund R, Gustafsson L, Niklasson C, Lidén G (1997) Ind Eng Chem Res 36:4659
59. Sjöström E (1993) Wood chemistry: fundamentals and applications. Academic Press, San Diego, CA
60. Hespell RB (1998) J Agric Food Chem 46:2615
61. Saddler JN, Ramos LP, Breuil C (1993) Steam pretreatment of lignocellulosic residues. In: Saddler JN (ed) Bioconversion of forest and agricultural plant residues. CAB International, Wallingford, p 73
62. Tengborg C, Stenberg K, Galbe M, Zacchi G, Larsson S, Palmqvist E, Hahn-Hägerdal B (1998) Appl Biochem Biotechnol 70/72:3
63. Takagi M, Abe S, Suzuki S, Emert GH, Yata N (1978) Production of alcohol directly from cellulose using cellulase and yeast. In: Ghose TK (ed) Bioconversion of cellulosic substances into energy, chemicals and microbial protein. IIT, Zürich, p 551
64. Stenberg K, Bollók M, Réczey K, Galbe M, Zacchi G (2000) Biotechnol Bioeng 68:204
65. Harman GE, Kubicek CP (1998) *Trichoderma* and *Gliocladium*. Taylor and Francis, London

66. Skipper N, Sutherland M, Davies RW, Kilburn D, Miller RC, Warren A, Wong R (1985) Science 230:958
67. Penttilä ME, Lars A, Saloheimo M, Lehtovaara P, Knowles JKC (1987) Yeast 3:175
68. Wong WKR, Curry C, Parekh RS, Parekch SR, Wayman M, Davies RW, Kilburn DG, Skopper N (1988) Biodegrad 6:713
69. González R, Ramón D, Perez-Gonzalez JA (1992) Appl Microbiol Biotechnol 38:370
70. Ooi T, Minomiguchi K, Kawaguchi T, Okada H, Murao S, Arai M (1994) Biosci Biotechnol Biochem 58:954
71. Van Rensburg P, Van Zyl WH, Pretorius IS (1998) Yeast 14:67
72. Penttilä ME, André L, Lehtovaara P, Bialey M, Teeri TT, Knowles JKC (1988) Gene 63:103
73. Van Rensburg P, Van Zyl WH, Pretorius IS (1996) Curr Genet 30:246
74. Takada G, Kawaguchi T, Sumitani J, Arai M (1998) Biosci Biotechnol Biochem 62:1615
75. Machida M, Ohtsuki I, Fukui S, Yamashita I (1988) Appl Environ Microbiol 54:3147
76. Mochizuki D, Miyahara K, Hirata D, Matsuzaki H, Hatano T, Fukui S, Miyakawa T (1994) J Ferment Bioeng 77:468
77. Adam AC, Rubio-Texeira M, Polaina J (1995) Yeast 11:395
78. Murai T, Ueda M, Kawaguchi T, Arai M, Tanaka A (1998) Appl Microbiol Biotechnol 64:4857
79. Petersen SH, Van Zyl WH, Pretorius IS (1998) Biotechnol Techn 12:615
80. Christgau S, Kauppinen S, Vind J, Kofod LV, Dalboge H (1994) Biochem Mol Biol Int 33:917
81. Stålbrand H, Saloheimo A, Vehmaanpera J, Henrissat B, Penttilä M (1995) Appl Environ Microbiol 61:1090
82. Margolles-Clark E, Tenkanen M, Luonteri E, Penttilä M (1996) Eur J Biochem 240:104
83. Setati ME, Ademark P, Van Zyl WH, Hahn-Hägerdal B, Stålbrand H (2001) Protein Expression Purif 21:105
84. Morosoli R, Zalce E, Moreau A, Durand S (1992) Secretion of xylanase from *Cryptococcus albidus* by *Saccharomyces cerevisiae* and *Pichia stipitis*. In: Visser J, Beldman G, Kusters-van Someren MA, Voragen AGJ (eds) Xylans and xylanases. Elsevier Science Publishers B.V., Amsterdam, The Netherlands, p 247
85. Crous JM, Pretorius IS, Van Zyl WH (1995) Curr Genet 28:467
86. Crous JM, Pretorius IS, Van Zyl WH (1996) Appl Microbiol Biotechnol 46:256
87. La Grange DC, Pretorius IS, Van Zyl WH (1996) Appl Environ Microbiol 62:1036
88. Li X-L, Ljungdahl LG (1996) Appl Environ Microbiol 62:209
89. Margolles-Clark E, Tenkanen M, Nakari-Setälä T, Penttilä M (1996) Appl Environ Microbiol 62:3840
90. Perez-Gonzalez JA, de Graaff LH, Visser J, Ramón D (1996) Appl Environ Microbiol 62:2179
91. La Grange DC, Pretorius IS, Van Zyl WH (1997) Appl Microbiol Biotechnol 47:262
92. Luttig M, Pretorius IS, Van Zyl WH (1997) Biotechnol Lett 19:411
93. La Grange DC, Claeyssens M, Pretorius IS, Van Zyl WH (2000) Appl Microbiol Biotechnol 54:195
94. Eliasson A, Christensson C, Wahlbom CF, Hahn-Hägerdal B (2000) Appl Environ Microbiol 66:3381
95. Moniruzzaman M, Dien BS, Skory CD, Chen ZD, Hespell RB, Ho NWY, Dale BE, Bothast RJ (1997) World J Microbiol Biotechnol 13:341
96. Bothast RJ, Nichols NN, Dien BS (1999) Biotechnol Prog 15:867
97. Johansson B, Christensson C, Hobley T, Hahn-Hägerdal B (2001) Appl Environ Microbiol (submitted)
98. Palmqvist E, Hahn-Hägerdal B (2000) Biores Technol 74:17
99. Palmqvist E, Hahn-Hägerdal B (2000) Biores Technol 74:25
100. Jönsson LJ, Palmqvist E, Nilvebrant NO, Hahn-Hägerdal B (1998) Appl Microbiol Biotechnol 49:691
101. Larsson S, Reimann A, Nilvebrant NO, Jönsson LJ (1999) Appl Biochem Biotechnol 77/79:91

102. Taherzadeh MJ, Gustafsson L, Niklasson C, Lidén G (1999) J Biosci Bioeng 87:169
103. Larsson S, Palmqvist E, Hahn-Hägerdal B, Tengborg C, Stenberg K, Zacchi G, Nilvebrant NO (1999) Enzyme Microb Technol 24:151
104. Palmqvist E, Almeida JS, Hahn-Hägerdal B (1999) Biotechnol Bioeng 62:447
105. Taherzadeh MJ, Gustafsson L, Niklasson C, Lidén G (2000) Appl Microbiol Biotechnol 53:701
106. Maiorella B, Blanch HW, Wilke CR (1983) Biotechnol Bioeng 25:103
107. Taherzadeh MJ, Niklasson C, Lidén G (1997) Chem Eng Sci 52:2653
108. Busturia A, Lagunas R (1986) J Gen Microbiol 132:379
109. Weusthuis RA, Pronk JT, van den Broek PJ, van Dijken JP (1994) Microbiol Rev 58:616
110. Meinander NQ, Boels I, Hahn-Hägerdal B (1999) Biores Technol 68:79
111. Lagunas R, Dominguez C, Busturia A, Saez MJ (1982) J Bacteriol 152:19
112. Oehlen LJ, Scholte ME, de Koning W, van Dam K (1994) Microbiology 140:1891
113. Teusink B, Diderich JA, Westerhoff HV, van Dam K, Walsh MC (1998) J Bacteriol 180:556
114. Diderich JA, Teusink B, Valkier J, Anjos J, Spencer-Martins I, van Dam K, Walsh MC (1999) Microbiology 145:3447
115. Sauer N, Stadler R (1993) Plant J 4:601
116. Prof N Sauer, Department of Plant Physiology, University of Erlangen, Germany (personal communication)
117. Kötter P, Amore R, Hollenberg CP, Ciriacy M (1990) Curr Genet 18:493
118. Rizzi M, Erlemann P, Bui-Thanh NA, Dellweg H (1988) Appl Microbiol Biotechnol 29:148
119. Lucas C, van Uden N (1986) Appl Microbiol Biotechnol 23:491
120. Kilian SG, Prior BA, du Preez JC (1993) World J Microbiol Biotechnol 9:357
121. Nobre A, Lucas C, Leao C (1999) Appl Environ Microbiol 65:3594
122. Does AL, Bisson LF (1989) Appl Environ Microbiol 55:159
123. Kilian SG, van Uden N (1988) Appl Microbiol Biotechnol 27:545
124. Weierstall T, Hollenberg CP, Boles E (1999) Mol Microbiol 31:871
125. Alcorn ME, Griffin CC (1978) Biochim Biophys Acta 510:361
126. Reifenberger E, Boles E, Ciriacy M (1997) Eur J Biochem 245:324
127. Diderich JA, Schepper M, van Hoek P, Luttik MA, van Dijken JP, Pronk JT, Klaassen P, Boelens HF, de Mattos MJ, van Dam K, Kruckeberg AL (1999) J Biol Chem 274:15,350
128. Coons DM, Boulton RB, Bisson LF (1995) J Bacteriol 177:3251
129. Reifenberger E, Freidel K, Ciriacy M (1995) Mol Microbiol 16:157
130. Rouhonen L, Aristidou A, Teleman A, Penttilä M (1998) Yeast as a Cell Factory - EC Framework IV Symposium. Vlaardingen, The Netherlands, p 173
131. Wieczorke R, Krampe S, Weierstall T, Freidel K, Hollenberg CP, Boles E (1999) FEBS Lett 464:123
132. Hochster RM, Watson RW (1954) Arch Biochem Biophys 48:120
133. Bruinenberg PM, de Bot PHM, van Dijken JP, Scheffers WA (1983) Eur J Appl Microbiol Biotechnol 18:287
134. Verduyn C, van Kleef R, Frank J, Schreuder H, van Dijken JP, Scheffers WA (1985) Biochem J 226:669
135. Neuhauser W, Haltrich D, Kulbe KD, Nidetzky B (1997) Biochem J 326:683
136. Verduyn C, Frank J, van Dijken JP, Scheffers WA (1985) FEMS Microbiol Lett 30:313
137. VanCauwenberge JE, Bolen PL, McCracken DA, Bothast RJ (1989) Enzyme Microb Technol 11:662
138. Bruinenberg PM, De Bot PHM, van Dijken JP, Scheffers WA (1984) Appl Microbiol Biotechnol 19:256
139. Rizzi M, Harwart K, Erlemann P, Bui-Thanh NA, Dellweg H (1989) J Ferment Bioeng 67:20
140. Rizzi M, Harwart K, Bui-Thanh NA, Dellweg H (1989) J Ferment Bioeng 67:25
141. Walfridsson M, Hallborn J, Penttilä M, Keränen S, Hahn-Hägerdal B (1995) Appl Environ Microbiol 61:4184
142. Walfridsson M, Anderlund M, Bao X, Hahn-Hägerdal B (1997) Appl Microbiol Biotechnol 48:218

143. Eliasson A (2000) PhD thesis, Lund University
144. Anderlund M, Rådström P, Hahn-Hägerdal B (2001) Metab Eng (in press)
145. Webb SR, Lee H (1992) J Gen Microbiol 138:1857
146. Zhang Y, Lee H (1997) FEMS Microbiol Lett 147:227
147. Kostrzynska M, Sopher CR, Lee H (1998) FEMS Microbiol Lett 159:107
148. Metzger MH, Hollenberg CP (1995) Eur J Biochem 228:50
149. Luyten K, Albertyn J, Skibbe WF, Prior BA, Ramos J, Thevelein JM, Hohmann S (1995) EMBO J 14:1360
150. Tamas MJ, Luyten K, Sutherland FC, Hernandez A, Albertyn J, Valadi H, Li H, Prior BA, Kilian SG, Ramos J, Gustafsson L, Thevelein JM, Hohmann S (1999) Mol Microbiol 31:1087
151. Prof SG Kilian, Department of Microbiology and Biochemistry, The University of the Orange Free State, South Africa (personal communication)
152. Bhosale SH, Rao MB, Deshpande VV (1996) Microbiol Rev 60:280
153. Kristo P, Saarelainen R, Fagerström R, Aho S, Korhola M (1996) Eur J Biochem 237:240
154. Krems B, Charizanis C, Entian KD (1995) Curr Genet 27:427
155. Dombek KM, Ingram LO (1987) Appl Environ Microbiol 53:1286
156. Träff K, Otero Cordero RR, van Zyl WH, Hahn-Hägerdal B: Genetically engineered yeast and mutants thereof for the efficient fermentation of lignocellulose hydrolysates. International patent application W0 99/54477 filed 20 April 1999
157. Yamanaka K (1969) Arch Biochem Biophys 131:502
158. Cordero Otero RR, Gárdonyi M, Hahn-Hägerdal B, van Zyl WH, Dackehag EAV: Xylose isomerase with improved properties. International patent application W00/61733 filed 7 April 2000
159. Ho NW, Chen Z, Brainard AP (1998) Appl Environ Microbiol 64:1852
160. Wahlbom CF, Eliasson A, Hahn-Hägerdal B (2001) Biotechnol Bioeng (in press)
161. Eliasson A, Boles E, Johansson B, Österberg M, Thevelein JM, Spencer-Martins I, Juhnke H, Hahn-Hägerdal B (2000) Appl Microbiol Biotechnol 53:376
162. Rodriguez-Pena JM, Cid VJ, Arroyo J, Nombela C (1998) FEMS Microbiol Lett 162:155
163. Teusink B, Walsh MC, van Dam K, Westerhoff HV (1998) Trends Biochem Sci 23:162
164. Boles E, Heinisch J, Zimmermann FK (1993) Yeast 9:761
165. Boles E, Lehnert W, Zimmermann FK (1993) Eur J Biochem 217:469
166. Hohmann S, Bell W, Neves MJ, Valckx D, Thevelein JM (1996) Mol Microbiol 20:981
167. Shi NQ, Jeffries TW (1998) Appl Microbiol Biotechnol 50:339
168. Skoog K, Jeppsson H, Hahn-Hägerdal B (1992) Appl Biochem Biotechnol 34/35:369
169. Jones RP (1989) Enzyme Microb Technol 11:130
170. D'Amore T, Panchal CJ, Russell I, Stewart GG (1990) Crit Rev Biotechnol 9:287
171. Brosnan MP, Donnelly D, James TC, Bond U (2000) J Appl Microbiol 88:746
172. Olsson L, Hahn-Hägerdal B (1993) Proc Biochem 28:249
173. Lindén T, Peetre J, Hahn-Hägerdal B (1992) Appl Environ Microbiol 58:1661
174. Hahn-Hägerdal B, Jeppsson H, Olsson L, Mohagheghi A (1994) Appl Microbiol Biotechnol 41:62
175. Rank GH, Casey G, Xiao W (1988) Food Biotechnol 2:1
176. Sakai K, Yamamoto M (1986) Agric Biol Chem 50:1177
177. Murata K, Fukuda Y, Shimosaka M, Watanabe K, Saikusa T, Kimura A (1985) Appl Environ Microbiol 50:1200
178. Zhu J, Contreras R, Fiers W (1986) Gene 50:225
179. del Pozo L, Abarca D, Claros MG, Jimenez A (1991) Curr Genet 19:353
180. Henderson RCA, Cox BS, Tubb R (1985) Curr Genet 9:133
181. Hinchliffe E, Daubney CJ (1986) J Am Soc Brew Chem 44:98
182. Hadfield C, Cashmore AM, Meacock PA (1987) Gene 52:59
183. Bussey H, Meaden P (1985) Curr Genet 9:285
184. Casey GP, Xiao W, Rank GH (1988) J Inst Brew 94:93
185. D'Amore T, Celotto G, Russell I, Stewart GG (1989) Enzyme Microb Technol 11:411
186. Gupthar AS (1992) Can J Microbiol 38:1233

187. Kohalmi SE, Kunz BA (1988) J Mol Biol 204:561
188. Lee GS, Savage EA, Ritzel RG, Von Borstel RC (1988) Mol Gen Genet 214:396
189. Tantirungkij M, Izuishi T, Seki T, Yoshida T (1994) Appl Microbiol Biotechnol 41:8
190. Nielsen J (1998) Biotechnol Bioeng 58:125
191. Vallino JJ, Stephanopoulos G (1993) Biotechnol Bioeng 41:633
192. Stephanopoulos G, Aristidou A, Nielsen J (1998) Metabolic engineering: principles and methodologies. Academic Press, San Diego
193. Marx A, de Graaf AA, Wiechert W, Eggeling L, Sahm H (1996) Biotechnol Bioeng 49:111
194. de Graaf AA, Striegel K, Wittig RM, Laufer B, Schmitz G, Wiechert W, Sprenger GA, Sahm H (1999) Arch Microbiol 171:371
195. Kacser H, Burns JA (1973) Symp Soc Exp Biol 27:65
196. Heinrich R, Rapoport TA (1974) Eur J Biochem 42:89
197. Galazzo JL, Bailey JE (1990) Enzyme Microb Technol 12:162
198. Ehlde M (1995) PhD thesis, Lund University
199. Mavrovouniotis ML (1993) Ismb 1:275
200. Nielsen J (1997) Biochem J 321:133
201. Lockhart DJ, Winzeler EA (2000) Nature 405:827
202. Pandey A, Mann M (2000) Nature 405:837

Received: December 2000

Metabolic Engineering of Indene Bioconversion in *Rhodococcus* sp.

Daniel E. Stafford, Kurt S. Yanagimachi, Gregory Stephanopoulos

Department of Chemical Engineering, Massachusetts Institute of Technology, Cambridge, MA 02139, USA, e-mail: gregstep@mit.edu

We have applied the methodology of metabolic engineering in the investigation of the enzymatic bioreaction network in *Rhodococcus* sp. that catalyzes the bioconversion of indene to (2*R*)-indandiol suitable for the synthesis of *cis*-1-amino-2-indanol, a precursor of the HIV protease inhibitor, Crixivan. A chemostat with a novel indene air delivery system was developed to facilitate the study of steady state physiology of *Rhodococcus* sp. I24. Prolonged cultivation of this organism in a continuous flow system led to the evolution of a mutant strain, designated KY1, with improved bioconversion properties, in particular a twofold increase in yield of (2*R*)-indandiol relative to I24. Induction studies with both strains indicated that KY1 lacked a toluene-inducible dioxygenase activity present in I24 and responsible for the formation of undesired byproducts. Flux analysis of indene bioconversion in KY1 performed using steady state metabolite balancing and labeling with [^{14}C]-tracers revealed that at least 94% of the indene is oxidized by a monooxygenase to indan oxide that is subsequently hydrolyzed to *trans*-(1*R*,2*R*)-indandiol and *cis*-(1*S*,2*R*)-indandiol. This analysis identified several targets in KY1 for increasing (2*R*)-indandiol product yield. Most promising among them is the selective hydrolysis of indan oxide to *trans*-(1*R*,2*R*)-indandiol through expression of an epoxide hydrolase or modification of culture conditions.

Keywords. Indene, Bioconversion, *Rhodococcus*, Crixivan, Flux analysis, Chemostat, Steady state, Radiolabeled tracers

1 **Importance of Biocatalysis in Pharmaceutical Manufacturing** . . . 86

2 **Microbial Indene Bioconversion** 89

2.1 Indene Bioconversion in *Pseudomonas* 89
2.2 Isolation of *Rhodococcus* sp. I24 and Characterization of Indene Bioconversion 89

3 **Systems for Metabolic Flux Analysis of Indene Bioconversion** . . . 90

4 **Metabolic Flux Analysis of Indene Bioconversion in *Rhodococcus* sp. KY1** 94

5 **Future Directions for Metabolic Engineering of Indene Bioconversion** 99

References 100

Advances in Biochemical Engineering/
Biotechnology, Vol. 73
Managing Editor: Th. Scheper

List of Abbreviations

$[C2R]$	*cis*-(1*S*,2*R*)-Indandiol concentration
$[C2R^*]$	[^{14}C]-*cis*-(1*S*,2*R*)-Indandiol concentration
$[C2S^*]$	[^{14}C]-*cis*-(1*R*,2*S*)-Indandiol concentration
$[I_{tot}]$	Indene concentration (labeled plus unlabeled)
$[I^*]$	[^{14}C]-Indene concentration
$[IO^*]$	[^{14}C]-Indan oxide concentration
$[K^*]$	[^{14}C]-1-Keto-2-hydroxy-indan concentration
K_m^i	Michaelis-Menten constant for enzyme *i*
k_{C2R}	Indan oxide hydrolysis rate constant to *cis*-(1*S*,2*R*)-indandiol
k_{DO}	Dioxygenase rate constant
k_i	Rate constant for enzyme *i*
k_{MO}	Monooxygenase rate constant
k_{RDH}	*cis*-(1*S*,2*R*)-Indandiol dehydrogenase rate constant
k_{SDH}	*cis*-(1*R*,2*S*)-Indandiol dehydrogenase rate constant
k_T	Indan oxide hydrolysis rate constant to *trans*-(1*R*,2*R*)-indandiol
k_{TDH}	*trans*-(1*R*,2*R*)-Indandiol dehydrogenase rate constant
$[M_{tot}]$	Total metabolite concentration (labeled plus unlabeled)
$[M^*]$	[^{14}C]-Metabolite concentration
r_{C2R}	*cis*-(1*S*,2*R*)-Indandiol excretion rate
r_{C2S}	*cis*-(1*R*,2*S*)-Indandiol excretion rate
r_{IND}	Indene uptake rate
r_{IO}	Indan oxide excretion rate
r_K	1-Keto-2-hydroxy-indan excretion rate
r_T	*trans*-(1*R*,2*R*)-Indandiol excretion rate
$[T^*]$	[^{14}C]-*trans*-(1*R*,2*R*)-Indandiol concentration
v_{C2R}	Indan oxide hydrolysis flux to *cis*-(1*S*,2*R*)-indandiol
v_{DO}	Dioxygenase flux
v_i	Flux for enzyme *i*
v_{max}	Maximum specific rate for an enzyme catalyzed reaction
v_{MO}	Monooxygenase flux
v_{RDH}	*cis*-(1*S*,2*R*)-Indandiol dehydrogenase flux
v_{SDH}	*cis*-(1*R*,2*S*)-Indandiol dehydrogenase flux
v_T	Indan oxide hydrolysis flux to *trans*-(1*R*,2*R*)-indandiol
v_{TDH}	*trans*-(1*R*,2*R*)-Indandiol dehydrogenase flux
X	Biomass concentration

1 Importance of Biocatalysis in Pharmaceutical Manufacturing

The design and use of small molecules against biological macromolecular targets is of considerable significance in the pharmaceutical industry. Increasingly important among them are chiral compounds whose therapeutic activity is due primarily to a single stereoisomer. These compounds accounted for 32% of worldwide drug sales in 1999 [1]. The selective activity of these drugs is a result of the differential binding characteristics particular stereoisomers have with the active site of a target enzyme. Different stereoisomers of a compound can have drastically reduced activity against a target or even toxic effects.

Protease inhibitors are well-characterized chiral drugs in terms of their mechanism of action. An important new class of protease inhibitors comprises molecules designed to treat HIV infection. In particular, indinavir sulfate (CRIXIVAN, Merck and Co., Inc.) contains five chiral centers that must be of a specific orientation for the molecule to have the desired therapeutic effect. Manufacturing processes for these compounds involving chemical synthesis steps can be quite inefficient, due to yield reduction caused by racemization at each step where a chiral center is formed. A key intermediate in the synthesis of CRIXIVAN is *cis*-(1*S*,2*R*)-1-amino-2-indanol [(–)-CAI], an indene derivative that contributes two chiral centers to indinavir sulfate (Fig. 1). To circumvent the technically demanding chemical synthesis of (–)-CAI and reduce product loss, Merck scientists conceptualized a bioconversion process in which indene is oxidized to one of three derivatives that can serve as precursors to (–)-CAI: *cis*-(1*S*,2*R*)-indandiol, *trans*-(1*R*,2*R*)-indandiol, or (1*S*,2*R*)-indan oxide. Oxygenases that have been identified in isolates of the genus *Pseudomonas* and *Rhodococcus* can catalyze this transformation.

Oxygenases are useful enzymes for introducing chiral centers to prochiral compounds in a stereospecific manner. These enzymes catalyze the initial step in the biodegradation of many aromatic compounds by a number of microorganisms. Oxygenases play a significant role in the metabolism of straight-chain alkyl and aromatic compounds, and also many halogen-substituted hydrocarbons [2, 3]. The broad applicability of these biocatalysts has generated strong interest in their function including the mechanisms of their activity and subunit composition of many oxygenases [4]. In general, the complex nature of these enzymes as well as their cofactor requirements necessitates the development of whole-cell bioconversion systems to prevent enzyme degradation and facilitate cofactor regeneration.

A number of oxygenases have been described to catalyze multiple transformations of indene. Toluene dioxygenase (TDO) has been found to possess both monooxygenase and dioxygenase activities. Wackett and co-workers induced *Pseudomonas putida* F39/D with toluene, which then converted indene to *cis*-(1*S*,2*R*)-indandiol in approximately 30% e.e. and (1*S*)-indenol in 26% e.e. [5]. Gibson et al. found in *Pseudomonas* sp. 9816-4 that indene serves as a substrate

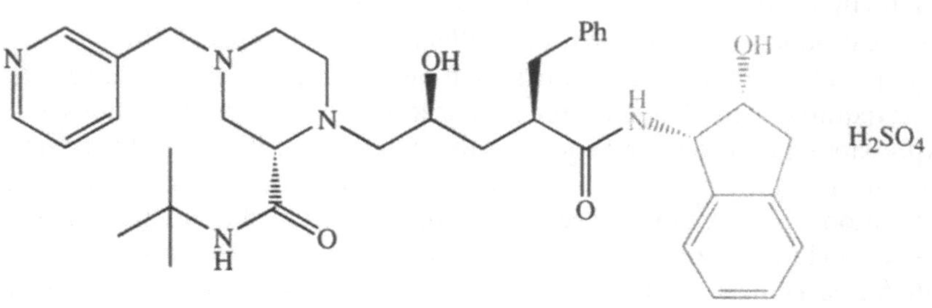

Fig. 1. Structure of indinavir sulfate. Shaded portion is *cis*-(1*S*)-amino-(2*R*)-indanol [(–)-CAI], and can be chemically synthesized from 1,2-indandiol of (2*R*) chirality

for both mono- and dioxygenation reactions by a naphthalene dioxygenase (NDO) to form (1*S*)-indenol and *cis*-(1*R*,2*S*)-indandiol, respectively [6]. The same products were detected in *Rhodococcus* sp. NCIMB 12038 when the cells were induced with naphthalene [7].

The quest for microorganisms capable of performing the desired biotransformation of indene led to the isolation of several strains of the genus *Rhodococcus* from soil samples contaminated with aromatic compounds that are able to oxidize indene to 1,2-indandiols of different chirality, and various other oxygenated derivatives [8]. Induction studies indicated that several oxygenases were present and differentially induced by naphthalene, toluene, and indene. The stereospecific nature of the enzymes expressed in *Rhodococcus* as well as their ability to tolerate indene as a substrate makes these microorganisms promising candidates for development as an industrial-scale biocatalyst for the production of (2*R*)-indandiol.

An effective whole-cell biocatalyst comprises a bioreaction network optimally configured for maximizing the yield and productivity of the desired product. The development of such a strain can best be performed within a metabolic engineering paradigm. This is based on a rigorous flux analysis that reveals the relative importance of the different metabolic pathways in the strain and suggests specific ways for further improvement. To enable the metabolic engineering of indene bioconversion in *Rhodococcus*, we had to develop (a) the prerequisites for flux analysis of relatively "uncharacterized" strains, such as those described here, where there is little a priori knowledge of the bioconversion pathways of interest, and (b) the tools for controlled and efficient genetic modification.

The foremost requirement is the accurate determination of an observable bioreaction network structure that describes indene oxidation in *Rhodococcus*. Based on product accumulation profiles and induction studies, indene bioconversion networks have been proposed for several isolates [8]. To validate further these networks for our strains and employ them for flux analysis we developed an experimental system that can maintain cells at steady state while allowing accurate metabolite measurements for flux determination. The system comprised a chemostat with a regular feed of liquid medium and separate supply of the indene precursor through a gas phase line. Indene uptake and metabolite production rates were easily measured in this system, leading to the calculation of the unknown bioconversion fluxes. Additional measurements for system closure and further validation were obtained by using radiolabeled tracers and measuring the products of their oxidation in *Rhodococcus* cultures [9, 10].

In parallel with the above efforts, we have also been developing the biological tools required to implement at the genetic level proposed gene deletions or overexpressions. For novel strains, such as the isolates described here, the genetics of bioconversion are relatively unknown and must be developed to allow implementation of any changes deemed appropriate from flux analysis. The tools needed include plasmids that can replicate in both *Rhodococcus* and *Escherichia coli*, for carrying a genomic library and manipulating cloned genes; selectable markers that must be determined for use in plasmids; and transformation methods to facilitate gene transfer between strains.

We review here our findings about the bioreaction network structure for indene bioconversion. We note that the indene bioconversion network in *Rhodococcus* is an isolated metabolic system for flux analysis because indene oxidation is significantly decoupled from primary metabolism since the *Rhodococcus* isolates of interest cannot utilize indene as a carbon source, although some cofactor requirements may be in common. The native functions of the oxygenase enzymes in the *Rhodococcus* networks are for toluene and/or naphthalene degradation as substrates. Use of glucose as a carbon source decouples the growth aspect of cell physiology from the bioconversion machinery of the cell. Growth-associated metabolism is a major source of uncertainties because of the many additional considerations it introduces into a metabolic engineering analysis. These bioconversion features distinguish our system from previous metabolic engineering applications.

2 Microbial Indene Bioconversion

2.1 Indene Bioconversion in *Pseudomonas*

A possible initial choice of strain to carry out indene bioconversion was *Pseudomonas putida,* which has been well characterized genetically and possesses a diverse metabolism of aromatic compounds. *Pseudomonas putida* F1 is known to express TDO capable of oxidizing indene to, among other products, *cis*-(1*S*,2*R*)-indandiol [5]. As this dioxygenase requires toluene for full induction, it was desired to remove toluene as a requirement to avoid substrate competition with indene for the dioxygenase [11]. Mutants were isolated that expressed TDO in the absence of toluene as an inducer, but TDO in these mutants exhibited poor stereospecificity. Enantiomerically pure *cis*-(1*S*,2*R*)-indandiol was obtained only at long culture times due to kinetic resolution catalyzed by a *cis*-(1*R*,2*S*)-indandiol dehydrogenase. Additionally, the yield of *cis*-indandiol from indene was low due to the monooxygenation of indene to 1-indenol (which isomerizes to 1-indanone in active cultures) by TDO. *cis*-Indandiol and the aforementioned co-oxidation products contribute to feedback inhibition of indene metabolism in *P. putida* [12]. To overcome these difficulties, a microbial screening program was undertaken to isolate strains able to tolerate both higher concentrations of indene and indene metabolites and dihydroxylate indene stereoselectively.

2.2 Isolation of *Rhodococcus* sp. I24 and Characterization of Indene Bioconversion

Rhodococcus sp. I24 was isolated from a toluene-contaminated aquifer and was found to oxidize indene to 1,2-indandiol and several other products. The undesired products 1-indenol and 1-indanone were formed directly from indene while racemic 1-keto-2-hydroxy-indan was formed from the indandiols. Based on product formation profiles and induction experiments, I24 was hypothesized

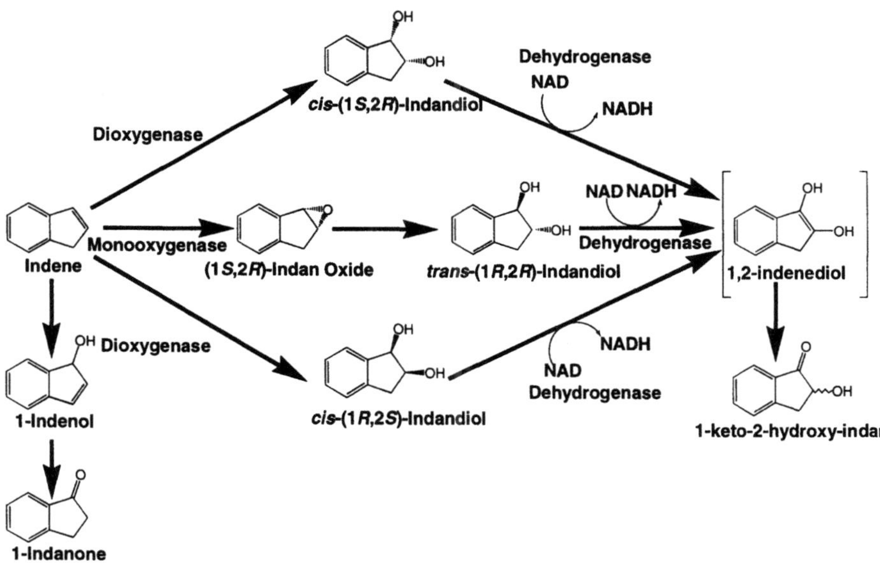

Fig. 2. Indene bioconversion network in *Rhodococcus* strain I24 proposed by Chartrain et al. [8]. Indene is converted to the indandiol enantiomers shown through specific oxygenase activities. Dioxygenases produce the *cis* enantiomers of indandiol while the monooxygenase converts indene to indan oxide. The indandiols are then converted to 1-keto-2-hydroxy-indan through the action of dehydrogenase enzymes and a proposed undetectable 1,2-indenediol intermediate

to contain a system of oxygenase enzymes that convert indene to various enantiomers of indandiol through the proposed bioreaction network shown in Fig. 2 [8]. The oxidation of indene to the indandiols followed by dehydrogenation is consistent with the degradation pathways elucidated for similarly structured compounds naphthalene and toluene in *Pseudomonas* [13]. However, the catechol analog 1-keto-2-hydroxy-indan is not oxidized via a ring-cleaving dioxygenase as has been determined in other aromatic degradative pathways. With indene as the sole aromatic compound present, I24 produced primarily *trans*-(1*R*,2*R*)-indandiol (>98% e.e.) in shake-flask cultures and withstood significantly higher concentrations of indene than *P. putida* in a two-liquid phase cultivation system that utilized silicon oil as an indene carrier [8]. Based on these findings, I24 emerged as a promising strain for subsequent development using a metabolic engineering approach.

3 Systems for Metabolic Flux Analysis of Indene Bioconversion

To improve the quantitative analysis of the indene bioconversion network, an experimental system enabling the accurate measurement of indene metabolites was developed. A multi-phase fermentation system commonly employed when dealing with substrates or products of relatively low solubility was not desirable for this analysis due to uncertainties associated with the partitioning of the indene metabolites between the aqueous and organic phases. In addition to the difficulty in obtaining a representative liquid phase sample to measure indene metabolite concentrations, the lack of partition coefficient data for indene metabolites made a single-phase system essential. To circumvent this issue, a continuous flow system that utilized a gas-phase delivery of indene was utilized [9]. The gas-phase concentration of indene was monitored using a photoionization detector and was manually controlled by mixing with a second air stream prior to sparging through the culture. By measuring the indene air concentration in the feed and exit gas streams of the chemostat, the indene uptake rate was calculated. In combination with the measurement of the liquid phase concentration of indene and other indene metabolites in the chemostat [8], the indene metabolite balances were closed (Table 1). Independent confirmation of the indene uptake rates calculated using the gas-phase indene concentrations was provided using [^{14}C]-indene uptake experiments, as will be described below.

Using this novel fermentation system, I24 was grown in a steady-state chemostat culture with an indene feed concentration of 85 ppm in 1.0 vvm of air, and a dilution rate of 0.10 h^{-1} [9]. In preliminary experiments with a continuous system, cell washout of I24 occurred when the indene concentration in the air feed exceeded approximately 200 ppm for dilution rates ranging from 0.05 h^{-1} to 0.10 h^{-1}. Thus, the indene feed utilized in the experiment described here is well under the toxicity limit of indene to I24. Upon reaching a steady state for five residence times, the primary indene metabolites detected were *cis*-indandiol, 1-keto-2-hydroxy-indan, and the undesirable byproducts 1-indenol and 1-indanone (Fig. 3). The lack of *trans*-indandiol formed may be due to a relatively high indene affinity of the dioxygenases relative to the monooxygenase under these culture conditions. When the indene feed concentration was increased from 85 ppm to 120 ppm with all other parameters held constant, a significant change in indene metabolism was observed after approximately ten residence times. Formation of 1-indenol and 1-indanone ceased, and the primary oxidation products were *trans*-indandiol, *cis*-indandiol, indan oxide, and 1-keto-2-hydroxy-indan. The yield of (2*R*)-indandiol from indene increased from approximately 30% to 60% following the metabolism shift (Table 1). The mutant with the altered metabolism from I24, denoted as strain KY1, was isolated from the chemostat and has shown indene metabolite profiles in steady-state and batch fermentations consistent with those observed following the metabolism shift in the I24 chemostat culture. Additionally, KY1 has been stable in numerous fed-batch experiments. It is believed that the KY1 strain evolved in response to the selective pressure applied by the chemostat environment to the I24 cells. The possibly toxic nature of 1-indenol and 1-indanone, especially at the high concentrations observed in the chemostat, facilitated the emergence of the new strain KY1 that is unable to oxidize indene to 1-indenol and 1-indanone. In steady-state chemostat studies performed with KY1, a substantially higher biomass concentration was obtained at a dilution rate of 0.065 h^{-1} than at 0.10 h^{-1} at an indene feed of 100 ppm, but the biomass concentrations were similar between the same two dilution rates at 170 ppm indene (Table 1). This may be a result of indene toxicity, the effects of which are presumably exerted more strongly at

Table 1. Steady state concentrations of *Rhodococcus* sp. KY1 chemostat cultures

Steady state values	$D = 0.10\ h^{-1}$		$D = 0.065\ h^{-1}$	
	100 ppm	170 ppm[d]	100 ppm	170 ppm
trans-(1*R*,2*R*)-Indandiol (mg/l)	86	181	151	262
cis-(1*R*,2*S*)-Indandiol (mg/l)	6	8	5	8
cis-(1*S*,2*R*)-Indandiol (mg/l)	24	52	35	55
1-Keto-2-hydroxy-indan (mg/l)	25	93	96	154
Indan oxide (mg/l)	21	42	34	55
Indene (mg/l)	10	14	5	6
Biomass (g DCW/l)	3.2	3.6	4.9	3.7
Indene uptake rate (material balance)[a,c]	35 ± 5	71 ± 5	29 ± 2	64 ± 5
Indene uptake rate (air measurement)[b,c]	40 ± 7	62 ± 12	28 ± 5	63 ± 10

[a] Determined from sum of indene metabolite excretion rates.
[b] Determined from inlet and outlet gas-phase indene concentrations.
[c] Uptake rates are in μmol/g DCW/h.
[d] Data for a pseudo-steady state when the concentrations were constant for one residence time.

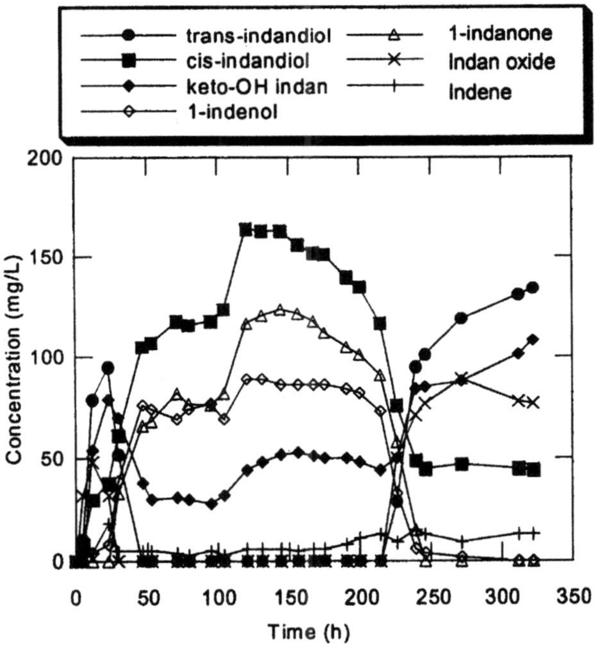

Fig. 3. Indene metabolite profiles in the *Rhodococcus* I24 chemostat at $0.10\ h^{-1}$ dilution rate. Indene was fed at 85 ppm from 0–105 h and subsequently at 120 ppm. Behavior characteristic of strain KY1 is exhibited after 250 h

higher feed concentrations. The higher metabolite concentrations observed for the 0.065 h^{-1}, 100 ppm state relative to the 0.065 h^{-1}, 170 ppm state suggests a possible correlation between biocatalyst concentration and indene metabolite titers. These data imply that an optimal fed-batch indene biotransformation be performed at relatively low indene feed to prevent growth attenuation due to substrate toxicity.

Induction studies that utilized [^{14}C]-indene as a probe were used to characterize more rigorously the indene bioconversion network of I24 and elucidate the difference(s) between the KY1 and I24 strains [9]. Cells were again grown in chemostat cultures in which naphthalene (40–70 ppm), toluene (100–200 ppm), or indene (100–110 ppm) was fed through the gas-phase until a steady-state was reached. The introduction of these compounds induced the activity of different oxygenases in the KY1 and I24 networks. Cells were removed from the chemostat culture and their physiology probed with [^{14}C]-indene. Specifically, by following the kinetics of formation of the primary oxygenated derivatives of [^{14}C]-indene following the introduction of the [^{14}C]-indene probe, the induction characteristics of key enzymes became apparent. Because of the rapid uptake of [^{14}C]-indene by the cells, the rate of tracer depletion was reaction-limited and provided a measure of the in vivo activity of these enzymes [10]. In cases where multiple enzymes were induced, oxygenase activity was estimated using the rate of formation of the appropriate [^{14}C]-indandiol.

Table 2 depicts the concentrations of [^{14}C]-labeled indene metabolites obtained after adding 25 µmol/l [^{14}C]-indene to I24 and KY1 cells under different inducers. These studies demonstrated that I24 expresses a toluene-inducible dioxygenase activity that produces primarily *cis*-(1*S*,2*R*)-indandiol and 1-indenol, and a naphthalene-inducible dioxygenase that produces primarily *cis*-(1*R*,2*S*)-indandiol and 1-indenol from indene. The tracer data also revealed that KY1 lacks the toluene-inducible dioxygenase present in I24 by virtue of the inability of KY1 to oxidize indene under toluene induction. The naphthalene-induced behavior of I24 and KY1 was similar. The slightly decreased excess of the *cis*-(1*S*,2*R*)-indandiol enantiomer produced by I24 relative to KY1 can be attributed to possible cross-induction of the toluene-inducible dioxygenase in I24.

Table 2. Conversion of 25 µmol/l [^{14}C]-indene to primary oxygenated products under different inducers after 5 min (reported as percentage of tracer added)

[^{14}C]-Metabolite	KY1			I24		
	Toluene	Naph-thalene	Indene	Toluene	Naph-thalene	Indene
cis-(1*R*,2*S*)-Indandiol	0	63	0	13	58	9
cis-(1*S*,2*R*)-Indandiol	0	0	0	26	7	18
1-Indenol	0	30	0	45	31	39
Indan oxide	0	0	16	0	0	0
Other	0	7	0	16	4	2
Indene (unoxidized)	100	0	84	0	0	32

Furthermore, tracer studies under indene induction showed that in KY1 the primary route of indene oxidation is through a novel monooxygenase activity to indan oxide presumed to be of (1*S*,2*R*) stereochemistry, while the metabolism in I24 closely resembled that observed under toluene induction.

Additional experiments were performed to confirm further the indene bioreaction network structures. Addition of indan oxide to both in vivo and cell-free systems showed that this intermediate is non-enzymatically hydrolyzed to both *trans*-(1*R*,2*R*)-indandiol and *cis*-(1*S*,2*R*)-indandiol in a 4:3 ratio, and that induction by indene had no effect on the hydrolysis rate [10]. A corollary of this result is that the *trans*-(1*S*,2*S*)- and *cis*-(1*R*,2*S*)-indandiol enantiomers are not formed by hydrolysis of (1*S*,2*R*)-indan oxide. This discounted the possibility that either (2*S*)-indandiol enantiomer is formed (from epoxide hydrolysis) but not detected due to rapid degradation to 1-keto-2-hydroxy-indan. Also, incubation of [^{14}C]-labeled *cis*-indandiols with induced I24 and KY1 cells resulted in only 1-keto-2-hydroxy-indan being formed, while [^{14}C]-*trans*-(1*R*,2*R*)-indandiol degradation was not detected in either strain. These data indicated that (a) there are no isomerization reactions occurring between the three indandiol enantiomers formed by indene oxidation in I24 and KY1, and (b) the dehydrogenase activity previously proposed to degrade *trans*-indandiol to 1-keto-2-hydroxy-indan is not present at a significant rate [10]. The latter conclusion is consistent with the observation by Chartrain et al. that *trans*-(1*R*,2*R*)-indandiol was dehydrogenated in I24 only at long culture times. In the context of a quantitative flux analysis of indene bioconversion, the flux supported by a *trans*-(1*R*,2*R*)-indandiol dehydrogenase was negligible relative to the flux through the other network reactions. Based on findings from these tracer studies, a new bioreaction network was proposed for the KY1 strain as shown in Fig. 4. The increased yield of (2*R*)-indandiol characteristic of KY1 made this the most interesting microorganism for further study using metabolic flux analysis.

4 Metabolic Flux Analysis of Indene Bioconversion in *Rhodococcus* sp. KY1

The indene bioconversion network proposed for *Rhodococcus* sp. KY1 (Fig. 4) using the induction studies with radiolabeled indene can be used to write five independent mass balances to describe six intracellular fluxes (Table 3). This yields an underdetermined system for the fluxes of the KY1 network requiring that at least one flux be directly measured to calculate uniquely the remaining network fluxes. It is further desirable to measure directly additional fluxes to generate redundancies that can be used to confirm the structure of the proposed bioreaction network, validate the flux estimates, and help detect gross measurement errors, if present.

[^{14}C]-*cis*-(1*S*,2*R*)-Indandiol was used to measure directly the corresponding steady state dehydrogenase flux, v_{RDH}, in the KY1 indene bioconversion network. By measuring the formation of [^{14}C]-1-keto-2-hydroxy-indan associated with the concomitant depletion of [^{14}C]-*cis*-(1*S*,2*R*)-indandiol in steady state cells, the

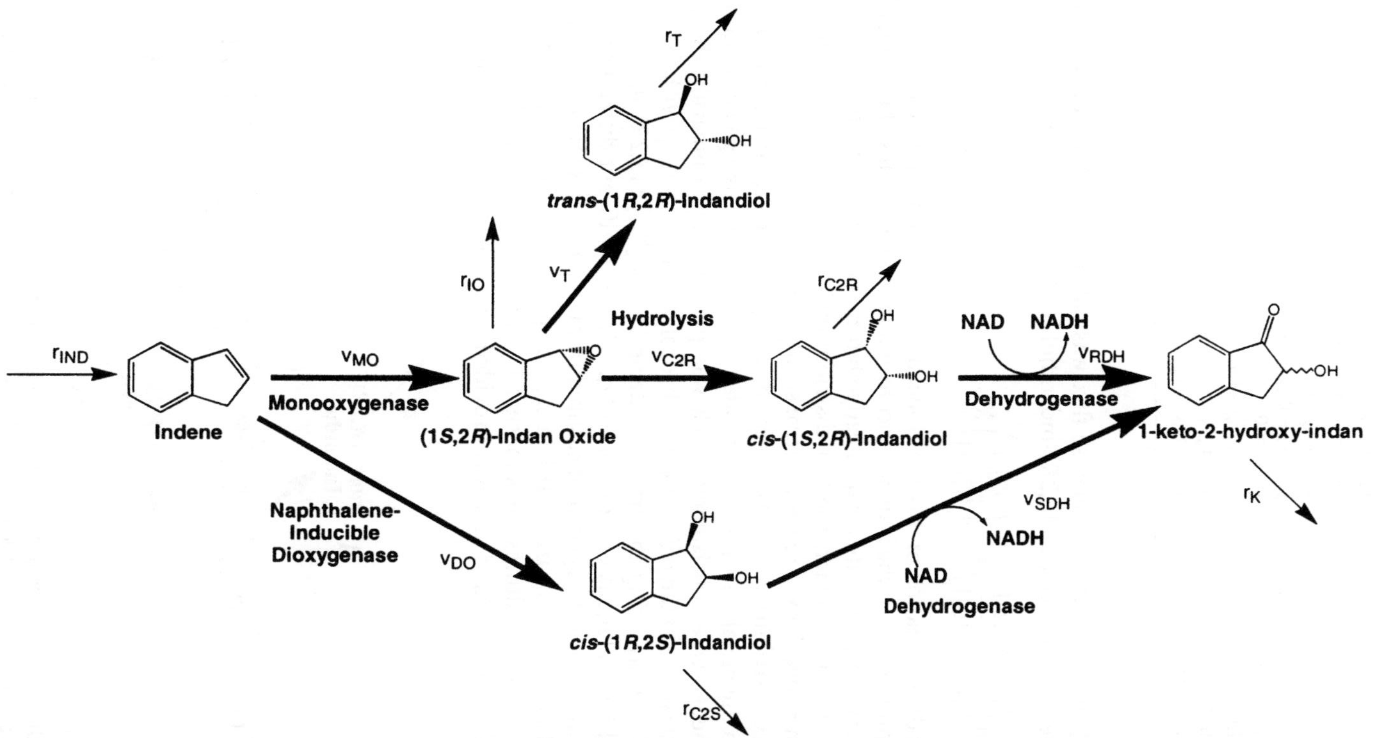

Fig. 4. Indene bioconversion network in *Rhodococcus* sp. KY1. Metabolite excretion and uptake rates are denoted by r_i, while intracellular fluxes are written as v_i

Table 3. Steady state metabolite balance equations for the KY1 indene bioconversion network

Metabolite	Mass Balance
Indene	$r_{IND} - v_{MO} - v_{DO} = 0$
Indan oxide	$v_{MO} - v_T - v_{C2R} - r_{IO} = 0$
trans-(1*R*,2*R*)-Indandiol	$v_T - r_T = 0$
cis-(1*S*,2*R*)-Indandiol	$v_{C2R} - v_{RDH} - r_{C2R} = 0$
cis-(1*R*,2*S*)-Indandiol	$v_{DO} - v_{SDH} - r_{C2S} = 0$
1-Keto-2-hydroxy-indan	$v_{RDH} + v_{SDH} - r_K = 0$

cis-(1*S*,2*R*)-indandiol dehydrogenase flux was calculated using Eq. (1), where $C2R^*$ is the radiolabeled *cis*-(1*S*,2*R*)-indandiol [$C2R$]:

$$\ln\left(\frac{[C2R^*\ (t)]}{[C2R^*\ (0)]}\right) = -\left(\frac{v_{RDH}}{[C2R]}\right)Xt \tag{1}$$

The direct determination of this additional flux (v_{RDH}) allowed the calculation of the remaining fluxes in the indene bioconversion network using the metabolite excretion rates (r_i) calculated from steady-state metabolite concentrations in the chemostat. Figure 5 depicts the flux distribution through the KY1 indene bioconversion network for a representative steady-state case [10].

The flux distribution results obtained using the directly determined v_{RDH} along with the metabolite production rates were validated using two redundant measurements. One consistency check was provided by comparing the predicted value of the indan oxide chemical hydrolysis ratio (v_T/v_{C2R}) with the value measured directly using the transient depletion of [^{14}C]-(1*S*,2*R*)-indan oxide tracer and an analogous expression to Eq. (1). No significant differences were found between the v_T/v_{C2R} ratios measured from the tracer experiment and that calculated as described earlier [10], in experiments where the tracer was added to both steady state cultures and supernatant and cell lysates of KY1 steady state cultures. A second redundancy was provided by comparing the indene uptake

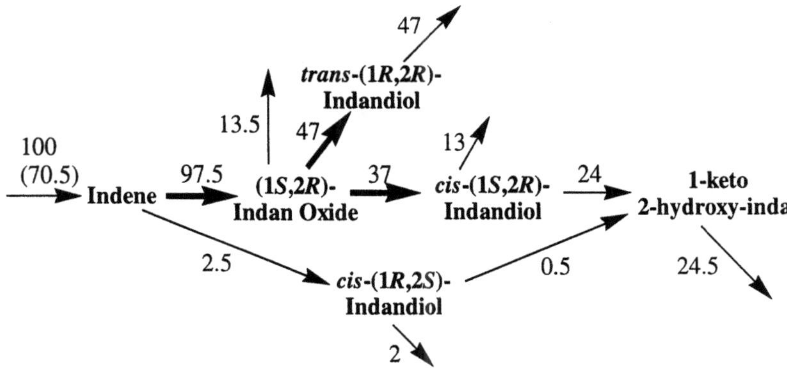

Fig. 5. Steady state intracellular flux distribution for KY1 at 100 ppm indene air feed concentration and a dilution rate of 0.065 h^{-1}. The fluxes were normalized by the indene uptake rate (*in parentheses*: µmol/h/g DCW)

rate calculated from the sum of the indene metabolite excretion rates with the indene uptake rate independently determined from the indene gas-phase concentrations in the chemostat. Both redundancy checks confirmed the flux estimates obtained from the metabolite balances and the direct measurement of v_{RDH}. Thus, any undetected perturbation of the steady state generated by the assaying procedure (i.e., alteration of NADH/NAD$^+$ ratios) was not significant enough to alter the measured flux distribution.

A final test of the intracellular fluxes determined by metabolite balancing was provided through comparison with the predictions of a first-order kinetic model describing the oxidation of pulsed [^{14}C]-indene to all detectable indene derivatives in steady state cells. Assuming Michaelis-Menten kinetics for a typical reaction depicted in Fig. 4, the rate of labeled metabolite conversion by that reaction can be expressed as

$$\frac{d[M^*]}{dt} = -\left[\frac{v^i_{max}[M_{tot}]}{K^i_m + [M_{tot}]}\right]\frac{[M^*]}{[M_{tot}]} = -\frac{v_i}{[M_{tot}]}[M^*] \tag{2}$$

If the concentration of M_{tot} remains constant in the course of the labeling experiment, the above expression is reduced to first-order kinetics with respect to the labeled metabolite concentration described by Eq. (3) below:

$$\frac{d[M^*]}{dt} = -k_i[M^*] \tag{3}$$

where

$$k_i = \frac{v_i}{[M_{tot}]}X \tag{4}$$

In all of the radiolabeled tracer experiments conducted, the concentrations of the corresponding indene metabolites were found to be constant so that the linear model with respect to the radiolabeled tracer is justified. However, for the mass balance on radiolabeled indene, the total metabolite concentration is not constant and the first-order kinetic model is only satisfied when the total concentration is sufficiently low such that $[M_{tot}] \ll K_m$ for that respective enzyme. Here, the flux is also not constant and can be expressed as shown in Eq. (5):

$$v_i = \frac{v^i_{max}}{K^i_m}[M_{tot}] \tag{5}$$

Substituting Eqs. (4) and (5) into the [^{14}C]-indene mass balance, the dynamics of [^{14}C]-indene depletion by all active oxygenases can be described using Eq. (6):

$$\frac{d[I^*]}{dt} = -\left[\sum_i\left(\frac{v^i_{max}}{K^i_m}\right)\right][I^*] \tag{6}$$

Thus, the dynamics of [^{14}C]-indene oxidation to downstream metabolites can be predicted using the flux estimates derived previously from metabolite balancing and direct flux measurement by translating these values into k_i estimates using Eqs. (4) and (5). These reaction rate constants can be used in the following equations that describe indene oxidation by KY1:

$$\frac{d[I^*]}{dt} = -(k_{MO} + k_{DO})[I^*] \tag{7}$$

$$\frac{d[IO^*]}{dt} = k_{MO}[I^*] - (k_{C2R} + k_T)[IO^*] \quad (8)$$

$$\frac{d[T^*]}{dt} = k_T[IO^*] - k_{TDH}[T^*] \quad (9)$$

$$\frac{d[C2R^*]}{dt} = k_{C2R}[IO^*] - k_{RDH}[C2R^*] \quad (10)$$

$$\frac{d[C2S^*]}{dt} = k_{DO}[I^*] - k_{SDH}[C2S^*] \quad (11)$$

$$\frac{d[K^*]}{dt} = k_{RDH}[C2R^*] + k_{SDH}[C2S^*] + k_{TDH}[T^*] \quad (12)$$

Figure 6 compares the experimentally measured metabolite profiles resulting from the oxidation of a pulse of [^{14}C]-indene by steady state chemostat cells with the kinetic profiles predicted by Eqs. (7)–(12) using flux values independently determined for the same steady state. The excellent agreement between the actual tracer data and the predicted oxidation profiles provides an additional validation of the fluxes calculated for the KY1 network.

Flux analysis of several steady states at different dilution rates and indene feed concentrations uniformly demonstrated that the key route of indene oxida-

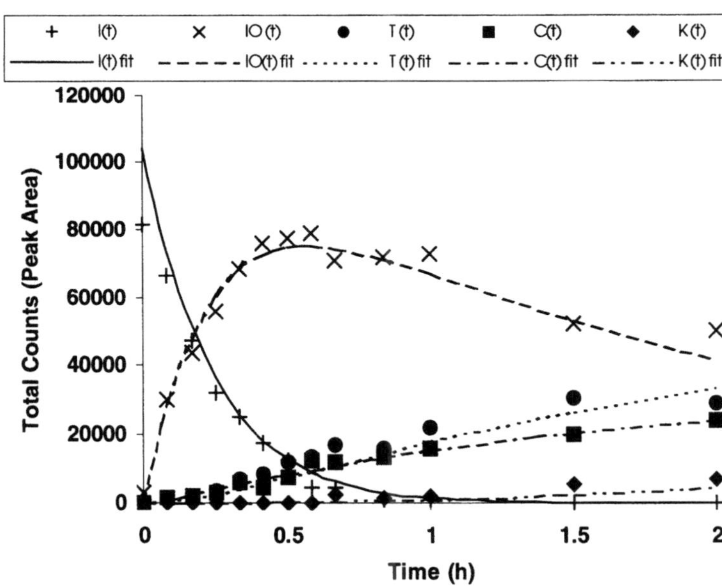

Fig. 6. Comparison of kinetic model predictions with experimental measurements of ^{14}C-indene metabolites for *Rhodococcus* KY1 cells obtained from a chemostat at steady state obtained with a dilution rate of 0.065 h^{-1} and 100 ppm indene air feed concentration. Reaction rate constants used in the kinetic model were determined from flux estimates as described in the text

tion in *Rhodococcus* sp. KY1 is through the novel monooxygenase enzyme. For all steady states analyzed, at least 94% of the indene was oxidized to indan oxide. This analysis also demonstrated that KY1 lacks a *trans*-(1*R*,2*R*)-indandiol dehydrogenase previously hypothesized to be present in the parent I24 strain. Additionally, the use of tracers showed a previously unidentified chemical step in the bioconversion network, namely the hydrolysis of indan oxide to *cis*-(1*S*,2*R*)-indandiol in addition to *trans*-(1*R*,2*R*)-indandiol.

5 Future Directions for Metabolic Engineering of Indene Bioconversion

A central finding of our analysis is that indene monooxygenase is the key enzyme for indene oxidation, and the most likely candidate for overexpression if further increase of the total oxidation flux of the indene network is desired. The emergence of indene monooxygenase as the main oxidizing enzyme in KY1 is contrary to the initial hypothesis that implicated toluene-induced dioxygenase as the main route for (2*R*)-indandiol biosynthesis. Estimates of monooxygenase activity in KY1 suggest that it is probably satisfactory for industrial-scale production. Assuming that indan oxide synthesis proceeds approximately at the same rate as indene depletion, a final titer of 8.7 g/l of product should be expected from a fed batch fermentation of three days duration at a cell density of 10 g/l. Other data indicating that *trans*-(1*R*,2*R*)-indandiol and 1-keto-2-hydroxy-indan may have an inhibitory effect on the monooxygenase, consistent with observations made in *P. putida* F1 [12], suggest this enzyme could also be considered as a candidate for directed evolution to reduce or eliminate product inhibition. Our revised view of the biocatalysis network emphasizes the need to express enzymes catalyzing the selective hydrolysis of indan oxide to *trans*-(1*R*,2*R*)-indandiol to prevent degradation by dehydrogenase(s). In terms of genetic modification, this task is more palatable than our original focus on multiple enzyme knockouts. Such secondary targets to improve (2*R*)-indandiol yield that were also identified by our analysis include the knockouts of multiple dehydrogenase activities and the dioxygenase producing *cis*-(1*R*,2*S*)-indandiol.

The presence of the *cis*-(1*S*,2*R*)-indandiol dehydrogenase means that the maximum yield of (2*R*)-indandiol that one can expect from KY1 is just under 60% due to the nature of the chemical hydrolysis of indan oxide to *trans*-(1*R*,2*R*)-indandiol and *cis*-(1*S*,2*R*)-indandiol. A promising approach to improving the product yield of KY1 is to hydrolyze selectively indan oxide to *trans*-(1*R*,2*R*)-indandiol by introducing an epoxide hydrolase and/or modifying culture conditions. A limonene-1,2-epoxide hydrolase from *Rhodococcus erythropolis* DCL14 has been characterized and cloned, and showed significant activity against indan oxide [14–16]. The activity of this enzyme encoded by the 0.5 kb *limA* gene should support the amount of indan oxide generated in KY1 by the indene monooxygenase. This would nullify the need for a dehydrogenase knockout since little or no *cis*-(1*S*,2*R*)-indandiol would be produced. Plasmids that can replicate in *Rhodococcus* were developed [17] that served as the foundation of a vector for the expression of this epoxide hydrolase in KY1, which has resulted in improved yield of *trans*-(1*R*,2*R*)-indandiol from indene [18]. Additionally, studies on the nature of indan oxide hydrolysis have shown that the ratio of *trans*-indandiol to *cis*-indandiol formed is highly pH-dependent. Further improvement of *trans*-indandiol yield has been obtained by performing the KY1 indene biotransformation at pH>8.0 [18].

Transaminase-type enzymes that can convert the indan oxide or (2*R*)-indandiols directly to (–)-CAI are also promising tools for the improvement of *Rhodococcus* as a biocatalyst. With the indandiols siphoned away to (–)-CAI, 1-keto-2-hydroxy-indan would not be formed and this product or *trans*-(1*R*,2*R*)-indandiol would not inhibit the monooxygenase enzyme activity. During KY1 fermentations, product inhibition of the monooxygenase by *trans*-indandiol and 1-keto-2-hydroxy-indan could also be avoided by removing the (2*R*)-indandiol product from the culture using resins or an organic phase. This technique has been applied to indene fermentations with *P. putida* using SP-207 resin to remove indandiols from unfiltered culture [12]. Additional factors that may contribute to the inhibition observed in fed-batch culture include the general toxicity of indene (and possibly other indene metabolites) to the cells, as well as the possible growth dependence of the expression of indene oxidation genes. These warrant further consideration as development with a viable production strain proceeds.

The metabolic engineering analysis of indene bioconversion in *Rhodococcus* species has been instrumental in defining ways to improve the strain and the fermentation process for the production of (2*R*)-indandiol. A pivotal event was the emergence of the KY1 strain that lacked competing dioxygenase activity and gave a higher product yield. This is believed to be a result of the application of selective pressure on the culture in a chemostat environment. This result supports a generic paradigm in this regard for evolution of strain properties in a properly designed continuous flow system.

Acknowledgements. This work was supported by a grant from Merck Research Laboratories. D. Stafford and K. Yanagimachi were supported in part by NIH Biotechnology Training Grant # 2T32 GM08334-10 and by the Engineering Research Program of BES, DoE Grant no. DE-FG02-94ER-14487.

References

1. Stinson S (2000) Chiral drugs. Chem Eng News 79:55–78
2. Finnerty W (1992) The biology and genetics of the genus *Rhodococcus*. Annu Rev Microbiol 46:193–218
3. Warhurst M, Fewson C (1994) Biotransformations catalyzed by the genus *Rhodococcus*. Crit Rev Biotechnol 14:29–73
4. Butler C, Mason J (1997) Structure-function analysis of the bacterial aromatic ring-hydroxylating dioxygenases. In: Advances in microbial physiology. Academic Press, vol 38, pp 47–84
5. Wackett L, Kwart L, Gibson D (1988) Benzylic monooxygenation catalyzed by toluene dioxygenase from *Pseudomonas putida*. Biochemistry 27:1360–1367
6. Gibson D, Resnick S, Lee K, Brand J, Torok D, Wackett L, Schocken M, Haigler B (1995) Desaturation, dioxygenation, and monooxygenation reactions catalyzed by naphthalene dioxygenase from *Pseudomonas* sp. strain 9816-4. J Bacteriol 177:2615–2621

7. Allen C, Boyd D, Larkin M, Reid KA, Sharma N, Wilson K (1997) Metabolism of naphthalene, 1-naphthol, indene, and indole by *Rhodococcus* sp. strain NCIMB 12038. Appl Environ Microbiol 63:151-155
8. Chartrain M, Jackey B, Taylor C, Sandford V, Gbewonyo K, Lister L, DiMichelle L, Hirsch C, Heimbuch B, Maxwell C, Pascoe D, Buckland B, Greasham R (1998) Bioconversion of indene to *cis*-(1*S*,2*R*)-indandiol and *trans*-(1*R*,2*R*)-indandiol by *Rhodococcus* species. J Fermentat Bioeng 86:550-558
9. Stafford DE, Yanagimachi KS, Lessard PA, Rijhwani SK, Sinskey AJ, Stephanopoulos G (2001) Optimizing bioconversion pathways through systems analysis and metabolic engineering (submitted)
10. Yanagimachi KS, Stafford DE, Dexter AF, Sinskey AJ, Drew SW, Stephanopoulos G (2001) Application of radiolabeled tracers to biocatalytic flux analysis (submitted)
11. Connors N, Chartrain M, Reddy J, Singhvi R, Patel Z, Olewinshi R, Salmon P, Wilson J, Greasham R (1997) Conversion of indene to *cis*-(1*S*),(2*R*)-indandiol by mutants of *Pseudomonas putida* F1. J Ind Microbiol Biotechnol 18:353-359
12. Buckland B, Drew S, Connors N, Chartrain M, Lee C, Salmon P, Gbewonyo K, Zhou W, Gailliot P, Singhvi R, Olewinshi R, Sun W-J, Reddy J, Zhang J, Jackey B, Taylor C, Goklen K, Junker B, Greasham R (1999) Microbial conversion of indene to indandiol: a key intermediate in the synthesis of CRIXIVAN. Metab Eng 1:63-74
13. Gibson D, Subramanian V (1984) Microbial degradation of aromatic hydrocarbons. In: Gibson D (ed) Microbial degradation of organic compounds. Marcel Dekker, New York, pp 253-294
14. Barbirato F, Verdoes J, de Bont J, van der Werf M (1998) The *Rhodococcus erythropolis* DCL14 limonene-1,2-epoxide hydrolase gene encodes an enzyme belonging to a novel class of epoxide hydrolases. FEBS Lett 438:293-296
15. van der Werf MJ, Overkamp KM, de Bont JAM (1998) Limonene-1,2-epoxide hydrolase from *Rhodococcus erythropolis* DCL14 belongs to a novel class of epoxide hydrolases. J Bacteriol 180:5052-5057
16. van der Werf M, Orru R, Overkamp K, Swarts H, Osprian I, Steinreiber A, de Bont J, Faber K (1999) Substrate specificity and stereospecificity of limonene-1,2-epoxide hydrolase from *Rhodococcus erythropolis* DCL14; an enzyme showing sequential and enantioconvergent substrate conversion. Appl Microbiol Biotechnol 52:380-385
17. Treadway S, Yanagimachi K, Lankenau E, Lessard P, Stephanopoulos G, Sinskey A (1999) Isolation and characterization of indene bioconversion genes from *Rhodococcus* strain I24. Appl Microbiol Biotechnol 51:786-793
18. Stafford DE, Yanagimachi KS, Lessard PA, Rijhwani SK, Sinskey AJ, Stephanopoulos G (2001) Optimizing bioconversion pathways through systems analysis and metabolic engineering (submitted)

Received: December 2000

Metabolic Engineering of the Morphology of *Aspergillus*

Mhairi McIntyre, Christian Müller, Jens Dynesen, Jens Nielsen

Center for Process Biotechnology, Department of Biotechnology, Building 223, Technical University of Denmark, 2800 Lyngby, Denmark, e-mail: jn@ibt.dtu.dk

The morphology of filamentous organisms in submerged cultivation is a subject of considerable interest, notably due to the influence of morphology on process productivity. The relationship between process parameters and morphology is complex: the interactions between process variables, productivity, rheology, and macro- and micro-morphology create difficulties in defining and separating cause and effect. Additionally, organism physiology contributes a further level of complexity which means that the desired morphology (for optimum process performance and productivity) is likely to be process specific. However, a number of studies with increasingly powerful image analysis systems have yielded valuable information on what these desirable morphologies are likely to be. In parallel, studies on a variety of morphological mutants means that information on the genes involved in morphology is beginning to emerge. Indeed, we are now beginning to understand how morphology may be controlled at the molecular level. Coupling this knowledge with the tools of molecular biology means that it is now possible to design and engineer the morphology of organisms for specific bioprocesses. Tailor making strains with defined morphologies represents a clear advantage in optimization of submerged bioprocesses with filamentous organisms.

Keywords. Morphological engineering, *Aspergillus*, Dimorphism

1 Introduction 104

2 Analysis Tools 105

3 Physiological Aspects of Morphological Development 105

3.1 Morphological Development of Filamentous Fungi 106
3.1.1 Apical Hyphal Extension 106
3.1.2 Cytoskeleton Organization 108
3.2 The Relationship Between Morphology and Productivity 109
3.2.1 Penicillin Production 111
3.2.2 Enzyme Production 112

4 Molecular Aspects of Morphological Control 114

4.1 Filamentous Fungi 114
4.1.1 Genes Involved in Morphology 114
4.1.2 Engineering Hyphal Architecture 116
4.2 Dimorphic Organisms 119
4.2.1 Biochemical Changes Associated with Dimorphism 120
4.2.2 Structural Changes Associated with Dimorphism 122
4.2.3 Molecular Level Control of Dimorphism 123

References 124

Advances in Biochemical Engineering/
Biotechnology, Vol. 73
Managing Editor: Th. Scheper

1 Introduction

Filamentous fungi are extensively used in the fermentation industry for the production of a long list of products including primary metabolites, antibiotics, industrial enzymes, and heterologous proteins. In the production of industrial enzymes, filamentous fungi are among the most important cell factories. This is due to their highly efficient secretion of proteins, and the establishment of good fermentation technology with these organisms. Protein secretion by filamentous organisms has been correlated with hyphal extension rates and tip growth and, as such, morphological characterization of the commonly used enzyme producing strains (mainly *Aspergilli*) is of interest. Additionally, fungal morphology is of interest due to the fact that it influences the rheology of the fermentation medium, and thereby has a significant impact on mixing and mass transfer within the bioreactor. In industry there is, therefore, a desire to tailor-make the morphology of filamentous fungi to ensure high protein secretion and at the same time a low viscosity culture.

Despite the importance of fungal morphology, our understanding of how the morphology can be manipulated is still rather limited. However, recent developments in basic biology have allowed progress in our understanding of fungal physiology and morphology by providing a number of morphological mutants and strains with disruption or inactivation of specific genes influencing the morphology. Studies employing such strains have greatly added to our knowledge of the regulation and control of morphology in filamentous fungi. A number of the key genes influencing morphology have been identified and it is, therefore, expected that in the future it will be possible to apply a much more directed approach to the development of better industrial strains.

To facilitate this process, this review will collate and summarize the current knowledge regarding fungal morphogenesis, with respect to both the physiological and molecular levels of control and regulation. The information on fungal physiology (growth and productivity) and morphology of filamentous fungi in submerged bioprocesses is relatively extensive compared to what is known about genetic control. In many cases, morphogenesis can be effected by changes in environmental conditions, while the molecular basis for such effects is not always known. On the other hand, morphological mutants have been identified, many with assumed "desirable" morphologies; however, the performance of these strains has not been assessed in submerged cultivation. Additionally, when considering tailoring morphologies for specific bioprocesses, here referred to as morphological engineering, it is not known which genes, either structural or regulatory, would be of interest.

Indeed, it is often the case that the link between control on the physiological level and the molecular basis for such control has not been made. In the past five to ten years, however, an increasing number of studies have identified genes involved in the control of morphology of filamentous fungi (namely *Aspergilli* and *Neurospora*). In addition, recent studies of dimorphic fungi have added further information on the genes involved in morphogenesis. The time is right, therefore, to begin building the picture of all factors known to influence morphology and discuss the possibilities for utilizing newly constructed strains for process optimization. This will provide a platform from which to push forward metabolic engineering of the morphology of all industrially relevant filamentous organisms.

2 Analysis Tools

The basis for rational design of fungal morphology is powerful analytical techniques. Computerized image analysis systems have been employed in studies of hyphal morphology for more than ten years [1–3] and have now reached a stage where reproducible analysis can be carried out (semi-) automatically and rapidly [4–6]. The resultant data can be used for studying growth mechanisms and kinetics and process modeling [7, 8] providing valuable information on the growth and differentiation of strains under different environmental conditions [9–11].

The application of fluorescent staining techniques to the study of filamentous organisms has provided valuable information on physiology, positioning of organelles and localization of structures within hyphae [12–15]. Indeed much has been learnt about the growth and organization of fungal hyphae through microscopy. When coupled with computerized image analysis, physiological information can be obtained in addition to the morphological data [7], providing two levels of detail on hyphal development.

Recently, studies employing a flow-through growth cell for analysis of the growth of filamentous fungi have been described [16, 17]. The system allowed the growth kinetics of single hyphae, from spore swelling and germination, to be determined on-line, rather than the average populations that are sampled from submerged bioprocesses. Clearly, such new advances and the application of "traditional" image analysis methods provide a valuable set of tools for studies of filamentous fungi, allowing quantification of changes resulting from metabolic engineering.

In addition, high performance bioreactors [18], particularly chemostats with, for example, Teflon coating to reduce wall growth can provide highly controlled environments for studies of morphologically engineered strains. Submerged cultivation under highly controlled conditions would be necessary to quantify precisely the effect of metabolic engineering of the morphology on productivity and bioreactor performance to allow accurate comparisons between strains.

3 Physiological Aspects of Morphological Development

Apical hyphal extension of filamentous fungi has been the subject of a number of thorough reviews [19–23] dealing with aspects of growth, hyphal architecture, and intracellular organization. For this reason, these subjects will not be discussed in detail here. Rather, the review of the physiology of fungal morphogenesis will focus on those features of hyphal development that may be of interest for designing strategies for the production of "better" industrial strains. With this aim in mind, particular focus will be on how the processes involved in apical hyphal extension are controlled and how this may be related to improved productivity.

3.1
Morphological Development of Filamentous Fungi

3.1.1
Apical Hyphal Extension

Fungal cells grow by apical hyphal extension in a highly polarized manner [14] with respect to their growth, morphology, organelle positioning, and cytoskeletal distributions [24, 25]. Hyphal extension is facilitated through deposition and insertion of new membrane and cell wall material at localized sites on the cell surface [21]. The enzymes and precursors required at the advancing tips for the synthesis of the new material are delivered in vesicles transported to these sites along a polarized cytoskeletal network [20, 21].

Figure 1 summarizes the processes involved in polar extension of filamentous organisms and the organization of the cell wall and cytoskeleton components. Supply of cell wall precursors is critical for wall expansion at the advancing tip and in many organisms the Spitzenkörper has been identified and visualized as the vesicle supply center [26–29]. This structure is likely also to have a role in controlling growth directionality [26]. The principal components of the cy-

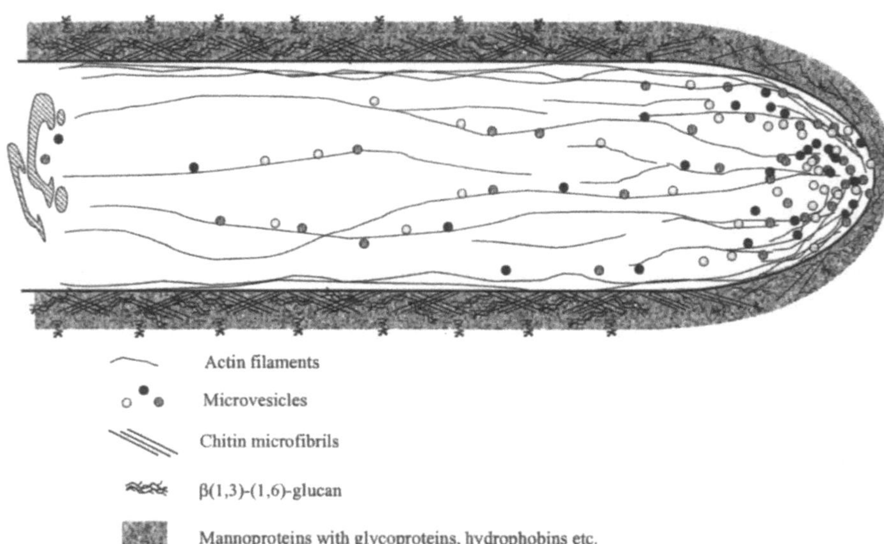

Fig. 1. Model of polar cell wall expansion in filamentous fungi. Vesicles with cell wall components and proteins are transported to the tip. An actin-myosin-based system is important in establishing polar growth through transport of the micro-vesicles to the cell surface. The cell wall at the apex is plastic but it hardens as the matrix of glucans and chitin crystallizes

toskeleton (actin and tubulin) have a major role in the process of tip growth, being responsible for the migration of organelles to the advancing apex [20,23,25]. While the regulation of polarity is complex and not fully understood, the role of Ca^{2+} ion gradients [30, 31], calcium mediated secondary messenger systems [32], and turgor pressure [19] have been demonstrated.

Biosynthesis of the cell wall material takes place in three sites, the cytoplasm, plasma membrane, and the wall itself. Deposition of cell wall components starts with several interconnected synthetic processes, which results in the extrusion of cell wall building blocks through the cellular membrane. Maturation of the wall through cross-linking of the components, then follows. The structural polymers chitin and $\beta(1-3)$ and $\beta(1-4)$ linked glucans contribute to the rigidity of the wall [33] and it is cross-linking of these that helps shape the hyphal architecture, adding a rigid structure to the mature wall. The enzymes involved in the cross-linking of chitin with wall components have not been identified, but it is most probable that transglycosidation leads to the formation of the cross-linkages. In filamentous fungi, autoradioactive studies following the incorporation of *N*-acetylglucosamine and glucose into growing hyphal walls have shown that nearly all *N*-acetylglucosamine is deposited within 1 μm of the hyphal tip region [34,35].

Fungi duplicate their length and nuclei through integration of the processes involved in tip growth, nuclear division, septation, and branching in a process termed the duplication cycle [36]. The duplication cycle in pre-divisional and post-divisional cells of *Aspergillus nidulans* is illustrated in Fig. 2. The cycle begins as a new apical compartment is created after septum formation has divided an existing apical compartment.

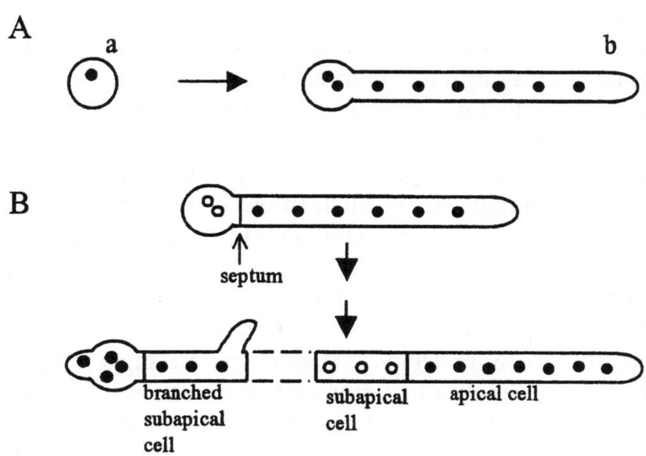

Fig. 2 A, B. Comparison of the duplication cycle and morphology of: **A** pre-divisional; **B** post-divisional cells of *A. nidulans*. A conidium (a) germinates and the first septum is formed at the basal end of the germ tube (b) when the germling has eight or more nuclei. Post-divisional cells are differentiated into subapical and apical tip cells (B). Apical cells contain many nuclei that are evenly spaced along the cell. Subapical cells contain three to four evenly spaced nuclei. Subapical cells can branch, and the branched cell grows like an apical cell. Apical and branched subapical cells have active nuclear cycles (*filled circles*) while nuclei in unbranched subapical cells are trapped in interphase (*empty circles*). (Revised from [37])

It has been argued that different fungal growth forms only differ in the degree of polarization of the processes involved in the formation of the new wall [38], with different types of fungal cells acquiring unique morphologies through distinctive patterns of polarized morphogenesis [39, 40]. For example, the ellipsoidal shape of yeasts occurs as a result of individual cells cycling through transient phases of polarized and isotropic growth. Conversely, filamentous organisms have cells (hyphae) that are long relative to their width. An understanding of how polarity is maintained, therefore, may provide an overview of how morphology, in general, may be manipulated through control of the processes leading to apical wall expansion.

Ultimately, polarized growth requires numerous gene products and coordination of processes involved in cytoskeleton and secretory functions [39]. At present we are still building information on how these events are coordinated and regulated. Although no complete picture of polarized apical growth exists, it is possible to study the effects of mutation on tip growth. Several genes have been identified whose products are involved in hyphal extension and mutant strains of filamentous fungi defective in polarity have been characterized. The possibility of morphological engineering via this route will be discussed in Sect. 4.

3.1.2
Cytoskeleton Organization

The fungal cytoskeleton is composed, principally, of two major polymers, microtubules and actin with a growing number of microtubule associated proteins (MAPs) and actin binding proteins (ABPs) being identified. The organized development of the cytoskeleton of filamentous fungi is crucial in shaping morphology, as it is the cytoskeleton that provides the scaffold for hyphal growth while, additionally, playing a role in directing polarity.

Actin has involvement in a variety of the processes that result in tip growth [25], and it is thought to play a multifunctional role in apical growth through the coordination of tip morphogenesis, cell wall synthesis, cytoplasmic migration, and organelle positioning [31, 41]. Filamentous actin (F-actin) is typically concentrated at the apices of filamentous fungi (Fig. 1), implying that it plays a role in tip extension [31]. The actin cap (the concentration of actin plaques located near the hyphal apex) appears to be responsible for tip extension and the actin cables (located subapically) are involved in the transport of vesicles to the extending tips. Studies of *Saprolegnia* [31], an Oomycete, suggest that the actin cap functions to support the apex in regions where the cell wall is weak, being optimally organized to reinforce the plastic cell wall at the growing tip. While the Oomycetes represent a different evolutionary line to *Aspergilli*, actin has been shown to have a primary role in the movement of secretory vesicles in fungi, and evidence for an actin-based system controlling polarity and secretion in *A. nidulans* has been presented [42].

It is likely that Ca^{2+} plays a role in controlling tip growth via actin in a number of diverse fungi. This is not only due to the fact that actin and Ca^{2+} are abundant in growing tips; Ca^{2+} ions are also known to regulate actin function in a number of ways. The subject has been extensively reviewed previously [31].

Calcium may also play a role as a branching signal, as has been investigated with *Neurospora crassa* [43], with the addition of the divalent cation ionophore inducing profuse branching. This observation has been linked to the involvement of cyclic AMP in the regulation of branching, as the colonial phenotype was dependent on a low intracellular level of cAMP, and there are known antagonistic regulatory roles of Ca^{2+} and cAMP [44]. Very little is known about how branching is regulated in filamentous fungi; however, a simple relationship between hyphal elongation rate and branch formation has been shown to exist in *A. nidulans* [45]. Branch initiation was observed in this organism when a compartment reached a maximum rate of extension, which was achieved at different lengths with different specific growth rates.

Further hyphal structure is provided through an arrangement of microtubules, formed through the polymerization of tubulin heterodimers. In addition to contributing to the internal scaffold of hyphal cells, these filaments have also been shown to be involved in the positioning of organelles in hyphae [43]. Nuclear migration plays an important role in the growth and development of filamentous fungi, as has been exemplified by studies on *A. nidulans* [46, 47]. Nuclear migration (and perhaps that of other organelles) is mediated by cytoplasmic dynein, a microtubule dependent motor [47, 48]. Actin related proteins, such as dynactin in *N. crassa* [49], have also been shown to be involved in the stabilization of the internal structure and the positioning of nuclei.

From the evidence presented above it appears that many of the components involved in shaping the hyphal ultrastructure have multifunctional roles, and this presents a complication if engineering of morphology is to proceed via regulation of structural genes. Multiple effects of structural gene inactivation are likely to be observed. Indeed, it would be most desirable if the phenotypes of strains with inactivated regulatory genes were to be investigated, and in any event that strains with single gene inactivations were characterized.

3.2
The Relationship Between Morphology and Productivity

A key aspect in metabolic engineering of *Aspergillus* morphology is the subsequent effect of morphology on product formation. Generally, morphological forms are described on two levels – macroscopic and microscopic [50, 51] with the macromorphology describing the gross morphology (pellets, clumps or freely dispersed mycelia) and the micromorphology describing the properties of these types (branch frequency, hyphal dimensions, and segregation, i.e., compartmentalization and physiological population distribution) [52, 53]. These descriptions are illustrated in Fig. 3. While the macroscopic morphology can influence medium rheology and thus mixing and mass transfer within a culture, the literature mainly describes control of macromorphology by environmental conditions. For example, *Aspergillus oryzae* produces pellets following spore agglomeration, a process which is pH dependent [53].

Figure 4 provides a schematic representation of the interactions between process conditions, morphology, and productivity. The micro-environment of hyphae is determined by the process conditions and the mixing of the culture, and it

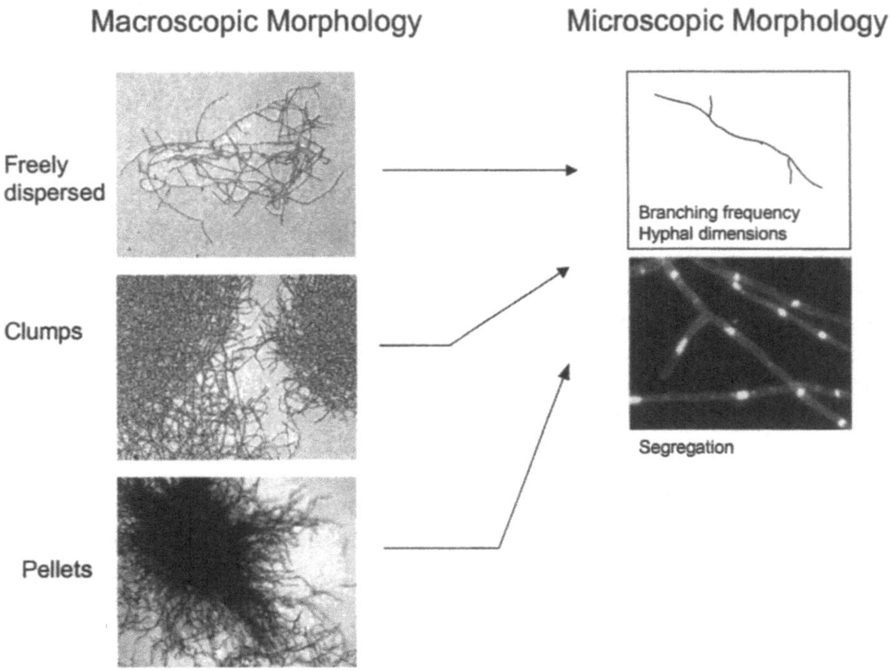

Fig. 3. Macroscopic and microscopic morphology of filamentous fungi. Macroscopic morphology describes the gross morphology, while microscopic morphology describes the properties (dimensions and compartmentalization) of the gross morphological forms

is the availability of nutrients and oxygen that determines the global regulation of genes. This in turn has influence on the genes directly controlling morphology or productivity. The resulting micromorphology can have a direct effect on metabolic pathway activity through the co-regulation of genes and can influence productivity due to the segregation of hyphae. Not all hyphal compartments are likely to have the same level of activity [7, 54]. Microscopic morphology also has other, indirect effects on productivity, with differentiation and hyphal dimensions influencing the secretion pathway. The processes of clumping and pelleting, and thus macromorphology, have significant influence on the measured mean activities or specific productivities of the cultures investigated [55]. Macroscopic morphology also determines the micro-environment of hyphae through effects on mixing, mass transfer, and culture rheology. Pellets may have dense and inactive cores due to poor diffusion of nutrients [51, 56], which may lead to cell lysis and thereby loss of the interior pellet structure [51]. Furthermore, the products of autolysis, which may be growth inhibitors, could diffuse through the pellets into the medium and inhibit the growth of the culture. Thus, development of macromorphologies indirectly affects the productivity of a culture.

If we are to consider metabolic engineering of the morphology of *Aspergilli*, or indeed filamentous fungi in general, efforts should be concentrated on understanding the processes that are represented within the shaded area on Fig. 4.

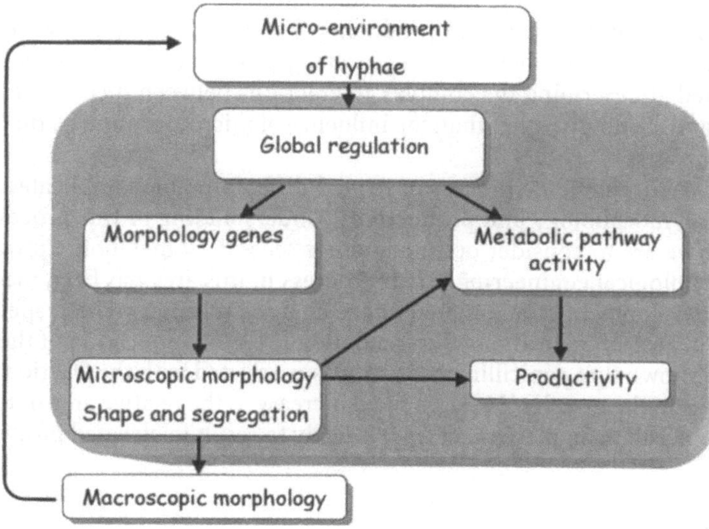

Fig. 4. Schematic representation of the interactions between process conditions, morphology, and productivity

It is only once the regulation and control of morphology is better understood that we can begin to engineer strains with better performance in submerged cultivation, with regard to productivity and physical properties of the culture.

Previous research has focused on the influence of morphology on either enzyme or secondary metabolite production, i.e., the major products from industrial bioprocesses utilizing filamentous organisms. With many advanced analysis tools in place (discussed in Sect. 2), detailed information on hyphal growth and kinetics can be obtained in a rapid and reproducible manner. Thus, effects of environmental changes or mutations on morphology can be quantified, allowing the relevance of these changes for process optimization to be assessed.

3.2.1 *Penicillin Production*

The effect of agitation on morphology and penicillin production by *Penicillium chrysogenum* has been the subject of a number of studies [55, 57–59]. Lower penicillin production was observed when agitation rates were high, a phenomenon which was attributed to the fact that mycelia were shorter and less branched. High agitation has been shown to promote rapid mycelial fragmentation [58, 59] and a higher branching frequency [58] for freely dispersed hyphal elements. Fragmentation of hyphal elements occurs when the local shearing forces become larger than the tensile strength of the cell wall [50]. The influence of the clumping of hyphae on rheology, and subsequently on penicillin production, has been widely discussed in the literature [55, 57–59]. The aim of many of the studies has been to optimize penicillin production by improved mixing, the resultant morphologies being quantified in an attempt to explain the results.

While these studies provide valuable information on the influence of environmental factors on the penicillin production process, they are of limited value from the viewpoint of metabolic engineering of morphology. This is because they aimed at describing the complex interactions between mechanical forces, growth, and rheology, rather than the influence of micromorphology on the production process.

In *P. chrysogenum*, the process of hyphal differentiation complicates studies correlating morphology and productivity. Understanding of this process is essential if we are to consider optimization of secondary metabolite production via morphological engineering. Great progress in this area has been made possible by development of automated image analysis routines written specifically for the purpose of quantifying differentiation [11, 60]. Application of these routines has shown that penicillin production is correlated with the fraction of subapical cells in the mycelia [50, 61], and an increase in the relative area of these regions (rather than an increase in tips) is likely to result in elevated productivity.

3.2.2
Enzyme Production

Protein secretion has been shown to occur at or very close to the tips of fungal hyphae [52, 62–64]. There have been a number of studies, therefore, attempting to correlate tip number with enzyme production [65, 66] and to investigate protein secretion by morphological mutants [52, 67, 68]. On investigating heterologous enzyme secretion by *Aspergillus niger* during continuous cultivations, Wongwicharn et al. [65] found that production was correlated with tip number as the concentration of oxygen was increased in the cultures. However, as metabolism, physiology (and thus protein secretion), and morphology are likely to be affected by the change in O_2 levels, such correlations should be treated with caution. The resultant changes in production may not be due to the changes in morphology alone; both physiology and morphology have been affected by the same external influence. Of more interest is the fact that these workers showed a further correlation between the active area (determined by biological staining) of hyphae and protein secretion, which is a more meaningful indication of the effect of increased oxygen in the influent gas. Fungal hyphae are not uniform, with respect to physiology, over their length [69]. Therefore, it is the observed changes in the active length, rather than overall length, of the hyphae that are likely to be responsible for alterations to growth and enzyme production [7].

Similarly, agitation rates [57, 58, 70] and biomass concentrations [55] are known to influence the physiological properties of a culture in addition to resulting in altered morphologies. Controlling such variables has been the strategy employed to alter morphology and investigate the subsequent effect on heterologous protein production in cultures of *Aspergillus awamori* [66]. Mean total hyphal length was found to decrease concomitant with increases in stirrer speed or increases in inoculum spore concentration. However, a reduced inoculum resulted in a more branched mycelium and an optimum stirrer speed was observed to result in a higher number of tips. In terms of productivity, the morphological differences had only a limited effect on product formation.

A clear picture of the effect of tip number on protein secretion is not apparent from the studies described above. Perhaps more insightful are the studies which have been carried out with morphological mutants, where comparisons of the effects of different morphologies may be more valid, being made without the influence of changes in environmental conditions. Spohr et al. [68] compared the α-amylase production in three strains of *A. oryzae* – a wild type, a transformed strain with an increased copy number of the α-amylase gene, and a morphological mutant of the transformed strain (which had a dense mycelium with more tips, relative to the other strains). The morphological mutant was found to be more efficient in producing α-amylase.

In a similar study [67], highly branched mutants of two strains of *A. oryzae* were investigated in submerged cultivation and morphology and protein secretion monitored. However, specific enzyme production was only improved in few of the highly branched strains, and the effect was dependent on the mode of cultivation. The authors concluded there was no clear correlation between branch frequency and the ability to secrete protein. The somewhat conflicting evidence presented above, concerning enzyme production in different morphological mutants, may be a result of the different types of morphological analysis applied in each of the studies. In general, only the freely dispersed (micromorphologies) were analyzed, and while these may be the predominant morphological form, they may not represent the total biomass. The relative amounts of the morphological forms are likely to be dependent on strain and cultivation conditions.

The observations from the studies above are further complicated by the fact that morphology also has an influence on broth rheology [71, 72] and thus can additionally affect production due to altered mixing and mass transfer in the culture fluid [55, 56, 73]. A linear relationship has been shown to exist between the degree of branching and the culture viscosity, with cultures of highly branched mutants being less viscous than wild type strains [67]. Mycelial morphology may not have a direct effect on protein secretion [70]; however, the relationship between agitation, morphology, and productivity must be considered when metabolic engineering of morphology is to be carried out. Changes in environmental conditions or mutant strains may appear to result in desirable morphological characteristics for improved productivity (e.g., increased number of tips). However, the performance of these strains in bioreactors remains to be the critical measure of their worth in process optimization.

From the evidence gathered, it is apparent that morphology has a significant role to play, influencing protein secretion either directly (tip number) or indirectly (by affecting mixing and mass transfer). Despite the conflicting results from submerged cultivations, direct evidence exists for protein secretion at the tips of fungal hyphae. Using immunogold labeling, Wösten et al. [64] localized secretion of glucoamylase in *A. niger* to the tips of actively growing hyphae. Further, staining with FITC conjugated antibody against α-amylase resulted in intense fluorescence of new tips and extending branches of *A. oryzae* [68]. Recently, visualization of proteins using GFP-fusions has allowed products of interest to be localized within hyphae [74, 75], providing additional evidence that protein secretion is an apical phenomenon. The importance of physiological information in addition to morphological data cannot be overstated when correlations between morphology and productivity are being formulated. In particular, fluorescence microscopy with biologically active stains has added greatly to our knowledge regarding the role of morphology in protein and secondary metabolite production. This is clearly an interesting route to pursue in morphological engineering.

4
Molecular Aspects of Morphological Control

4.1
Filamentous Fungi

Many genes have been identified in filamentous fungi where deletion or disruption results in morphological aberration. In many cases the gene product has not been identified, and in other cases has been shown to be a protein with a regulatory function. In fewer cases, the gene has been cloned, the function of the protein identified, and the morphological phenotype after disruption/deletion of the gene has been fully characterized. In the fewest of cases the organism has been studied in submerged cultivation and perhaps the effect of the morphological defect on productivity has been examined.

In the following section we have attempted to give an overview of the genes involved in controlling morphology in *Aspergilli*, illustrating with examples from the fewest studies where a clearer picture of the role genes in morphological development is available. It is using these examples that allows discussion of metabolic engineering of morphology and where we can begin to relate genetic manipulation of morphology genes to bioreactor performance and productivity.

4.1.1
Genes Involved in Morphology

Table 1 lists the genes of *A. nidulans* that have been identified as having a role in morphology, where the role of the protein encoded is known. The functions of the proteins listed are mainly related to the establishment and maintenance of hyphal polarity with inactivation of the corresponding genes resulting in swollen hyphae or aberrant branching patterns. Clearly, the interest is in exploiting this information for the improvement of submerged bioprocesses. It may be desirable to obtain a homogenous culture of filamentous fungal cells where polarity has been lost, thus leading to a culture giving rise to a lower medium viscosity and thereby an improved mixing of the culture, compared to a truly filamentous culture (see also Fig. 4). On the other hand, a culture which is hyperbranched may be desirable for the production of heterologous proteins where increased tip number may result in increased secretion and improved yields. Certainly, what still remains to be determined is the performance of many such morphological mutants in submerged culture with respect to growth and production characteristics

Protein kinases have proven to belong to ever-expanding gene/protein families and some of these have been shown to be very important in directing the tip

Table 1. Genes involved in morphological development of *Aspergilli* and subsequent effect on morphology following gene disruption

Gene	Protein function	Morphology obtained on gene inactivation	Reference
hypA/podA	Establishment and maintenance of hyphal polarity. Activation of growth arrest in subapical cells	Wide hyphae with thick lateral cell walls. High frequency of dichotomous apical branching	37, 39
hypC	Cell size control and control of spacing of septa.	Short subapical cells. High branching frequency	37
podB	Establishment and maintenance of hyphal polarity. Required for cytoskeletal organization in tip cells	Swollen hyphae	39
sepA	Formin. Control and organization of actin filaments at sites of localized cell wall deposition	Aseptated, wide hyphae. High frequency of dichotomous apical branching.	76
swoA	Maintenance of hyphal polarity	Swollen hyphae	77
swoF	Establishment and maintenance of hyphal polarity	Swollen hyphae	77

extension of hyphal cells. Protein kinases mediate the phosphorylation that regulates protein function directly, or via signal transduction, in many areas of the cell metabolism. The analysis of protein kinases in filamentous fungi is still in its early stages; however, it has already become clear that protein kinases are essential in linking signal transduction cascades, protein modification, and fungal morphogenesis. In *Saccharomyces cerevisiae* computer-based sequence analysis of the genome has revealed 113 genes which can be identified as protein kinases [78]. In filamentous fungi, and in eucaryotes in general, the protein kinases that phosphorylate either serine or threonine (Ser/Thr kinases) represent virtually all of the kinases described and this group includes cAMP-dependent kinases (PKA), protein kinase type C (PKC), mitogen-activated kinases (MAP), and p21-activated kinases (PAK) [79].

Of these, PKAs seem to have an important role in fungal development. However, caution should be exercised about specific function since no direct substrates for PKA have been identified yet [79]. In the plant smut fungus *Ustilago maydis*, cAMP signaling controls the dimorphic switch between the budding yeast form and (virulent) filamentous growth and it is also known to be involved in virulence of the rice blast fungus *Magnaporthe grisea* [80].

From an industrial perspective, an interesting study was carried out in *N. crassa* with the temperature-sensitive *mcb* mutant, which has a mutation in a regulatory subunit of the cAMP dependent protein kinase A. The strain displayed a complete loss in growth polarity at the restrictive temperature [81] and also in minimal medium supplemented with carboxymethyl cellulose (CMC) and sucrose [82]. This resulted in a considerable increase in the growing surface area of the fungus. It was hypothesized that protein secretion was limited by the amount of growing surface area; the protein secretion of the *mcb* mutant in liquid medium had a threefold higher yield of extracellular protein on biomass than the wild-type (50 mg/l to 15 mg/l). In addition, in the supplemented medium the yield of units CMCase on biomass was 20-fold higher. CMCase is mainly produced late in the cultivation and, therefore, it was stated that the level of protein production was not likely to be linked with the hyphal growth rate. However, hyphal growth rate was not measured, and it is likely that the CMCase production might be induced only when sucrose is depleted and as a result of the growth kinetics determined by the medium. (The wild-type grows fast to a high biomass concentration and experiences sucrose depletion more suddenly than the *mcb* mutant.) Therefore, it might have stopped its growth before it could produce the necessary proteins for CMCase production. CMCase production is complex to examine and, in addition, the protein secretion capacity of *N. crassa* is very low compared to the levels of *Trichoderma* or *Aspergilli* (g/l). As such, it may be very interesting to examine the effect of an *mcb* mutation on industrial-level protein producing strains of these species.

4.1.2
Engineering Hyphal Architecture

As discussed in Sect. 3, chitin synthesis is important in determining fungal cell shape and this process, in combination with embedding of polymers in the cell wall, is central in determining tip growth, branching, and differentiation of cell walls. For these reasons, the chitin synthases of *A. nidulans* and *Aspergillus. fumigatus* have been studied in some detail as targets for antifungal drugs. Additionally, *Aspergillus* strains disrupted in one or more chitin synthases have been shown to have altered morphologies and, therefore, it may be possible to regulate morphology by genetic manipulation of chitin synthases.

Chitin synthases catalyze the polymerization of *N*-acetylglucosamine (NAG) residues linked by $\beta(1-4)$ glycosidic bonds. The product is chitin, which is an unbranched polysaccharide that in fungi is aggregated into microfibrils with hydrogen bonds cross-linking adjacent chains [83]. In yeast, the chain length has been reported to be about 100 residues [84]. The microfibrils are located at the innermost part of the fungal cell walls where they exist as a rigid three-dimensional web capable of retaining its shape even when the matrix materials in which it is embedded are removed [85].

In *A. nidulans*, four chitin synthases have been cloned (*chsA*, *chsB*, *chsC*, *chsD*) as well as a gene, *csmA*, encoding a chitin synthase with a myosin motor-like domain fused at the N-terminus [86]. These chitin synthase genes are classified as class II, III, I, IV, and V, respectively, according to the amino acid similarity system of Bowen et al. [87]. Classes III and V of chitin synthases have been found exclusively in filamentous fungi, signifying a need for chitin synthases with specialized functions, perhaps because of the diversity of the processes requiring chitin deposition. The sites of chitin synthesis in *A. nidulans* are shown in Fig. 5, which also indicates where the gene products are most active.

Systematic studies with *A. nidulans* have shown, that the gene products of *chsA*, *chsC*, and *chsD* are involved in conidiophore formation (conidiation) and consequently spore production [88] (Fig. 5). Double mutants with *chsA*/*chsC* and *chsA*/*chsD* disruptions severely reduce spore production, signifying that the genes have functional overlap, but surprisingly, no effect was found in a *chsC*/*chsD* disruption. This points to the fact that *chsA* plays a main role in conidiation while *chsC* and *chsD* might be supplementary enzymes for two different parts of the conidiation. The two other known chitin synthases, *chsB* and *csmA*, are also important in spore production, signifying that all chitin synthases are involved in the complex conidiation process. However, Borgia et al. [89] found it probable (based on heterocaryon studies) that *chsB* does not take part in synthesis of the conidia itself. In the case of *csmA* disrupted strains, Horiuchi et al. [90] observed short stalks on the conidiophore vesicle, indicating a role for this chitin synthase in organizing the conidiophore vesicle.

Little is known about the in vivo regulation of chitin synthases in filamentous fungi. Spatial regulation requires either a mechanism for proper targeting of the active chitin synthase and/or a strictly localized activation of random dispersed chitin synthases at the site where chitin synthesis is required. In yeast, localization and activation of chitin synthases are affected not only by ions, metabolites, and zymogenicity but, as has been demonstrated with CHS3, by a large number of proteins such as activator proteins, translocational proteins, and septins [91] and perhaps also phosphorylation [92]. It seems that chitin synthase activity is regulated in a similar complex manner in filamentous fungi, for example, in both yeast and *A. fumigatus* the major part of chitin is synthesized by a non-zymogenic form [93].

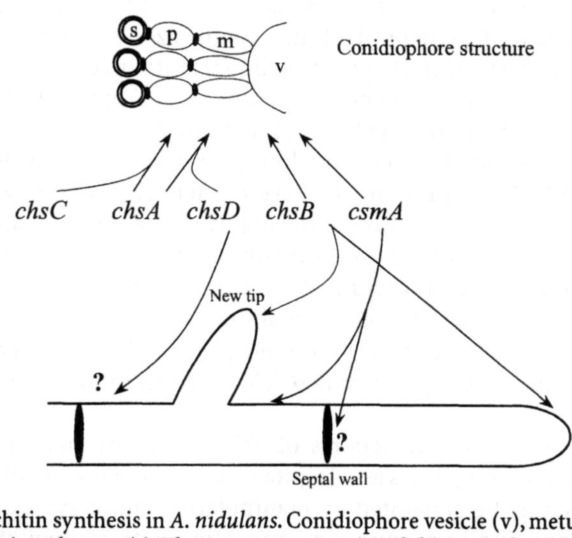

Fig. 5. Sites of chitin synthesis in *A. nidulans*. Conidiophore vesicle (v), metulae or sterigmata (m), phialides (p), and spore (s). The *arrows* suggest site of chitin synthesis based on observed mutant phenotypes. The products of the genes *chsA*, *chsC*, *chsA*, and *chsD* seem to have functional overlap

The formation of new branches requires considerably localized chitinase and glucanase activity, which must be both directed and activated precisely. Regulation of chitin synthase activity has been postulated to occur in the following way. Chitinases, located in lysosomal vesicles [94], may be released through the plasma membrane to the cell wall, lysing the chitin present there. This in turn could be broken down by *N*-acetyl glucosaminidase to yield *N*-acetyl glucosamine, which may activate the local chitin synthases [95] in the new tip. However, this mechanism has yet to be verified in vivo.

The effects of chitin synthase gene inactivation are summarized in Table 2. In *A. nidulans* disruptants of *chsA*, *chsC*, and *chsD* there are no phenotypic changes reported during hyphal growth although a *chsD* disruptant has been reported to have reduced cell wall chitin [96]. However, *chsA*/*chsC* double mutants were sensitive to salts, SDS, the chitin-binding dyes Calcofluor White and Congo red, and chitin synthase inhibitors [88] indicating ill-defined roles for all three chitin synthases in hyphal growth. In *csmA* disruptions it was found [90] that septa were irregularly positioned, a trait that was remedied when the full gene (including the myosin motor) was expressed driven by the *alcA* promoter but not when only the chitin synthase part of *csmA* was expressed. This indicates that the myosin motor domain is important for spatial regulation of this chitin synthase and for septum formation. The *csmA* disruption also displayed swelling of older parts of the hyphal cell walls, abnormal conidiophores, hypersensitivity to Calcofluor White, and low chitin content [96], indicating a general interference in chitin synthesis in the strain. Therefore, CSMA seems to have a role in maintaining hyphal cell wall integrity and establishing polarized cell wall (or septal) synthesis.

The *chsB* mutant of *A. nidulans* had a very reduced specific growth rate and produced stunted and bulging, highly branched hyphae suggesting that the *chsB* product is very important for the synthesis of chitin at the apical tips in *A. nidulans* [89, 100]. The *chsB* gene product only synthesizes a minor chitin sub-fraction (Table 2) but it has been shown to be important for correct organization of the hyphal growth. Studies using heterocaryons show that the *chsB* gene product is not readily diffusible in the hyphae and that individual chitin synthase molecules act in areas of the mycelium in close proximity to the nucleus encoding the molecule [89]. In contrast to the severe phenotype observed in the *chsB* mutant of *A. nidulans*, a disruption of the highly similar (88.9% similarity) *chsG* mutant of *A. fumigatus* was not as severely inhibiting to growth. The hyphae were hyper-branched but not stunted or bulging [93], indicating that other chitin synthases are capable of maintaining well-organized polar growth. Interestingly, it does not seem to be the other class III chitin synthase (chsC) of *A. fumigatus* since disruption of *chsC*/*chsG* had the same effect as the *chsG* mutant.

So far there have been no reports of chitin synthase manipulated strains grown in submerged bioprocesses, despite evidence suggesting direct morphological changes may be generated by manipulating chitin synthases. This may make them interesting to examine in connection with fermentation rheology and product secretion. It might be possible that the increased number of tips seen in the *chsB* mutant could enhance enzyme secretion or that other chitin

Table 2. Phenotypic effect of single and double chitin synthase gene inactivation in *A. nidulans*. The nomenclature of Horiuchi et al. [90] has been used as opposed to that of Specht et al. [96]. For clarity chsD [90] = chsE [96] and csmA [90] = chsD [96]

Chitin	Effect of gene inactivation		Reference
	On hyphal growth	On chitin content and conidia formation	
chsA	No observed effect	10% decrease in chitin content 30–40% loss in conidia formation	97 98
chsC	No observed effect	No observed effect	99
chsD	No observed effect	No effect on chitin content 45% loss of conidia formation No loss in conidia formation 30–40% decrease in chitin content.	99 97 96
chsB	Stunted and bulging highly branched hyphae	No effect on chitin content Reduced (55%) conidia formation	89 100
csmA	Intrahyphal hyphae and disturbance of septation	Swollen conidiophore vesicles	90
	Ballooned cell walls at subapical regions	40% decrease in chitin content 80% loss in conidia formation	96
chsA and *chsC*	No observed effect	Conidia formation almost totally lost (99.9%)	88
chsA and *chsD*	No observed effect	~30% decrease in chitin content 90%–97% loss in conidia formation	99 97
chsC and *chsD*	No observed effect	Same effect as in *chsD* inactivated strain	99
csmA and *chsD*	Same effect as in *csmA* inactivated strain	Same effect as in *csmA* inactivated strain	90

synthases could be manipulated, making the hyphal structure in such a way that the viscosity of the fermentation culture may be lowered.

4.2 Dimorphic Organisms

The study of dimorphic organisms is extremely relevant when considering the factors controlling and regulating morphology, particularly as investigations of these organisms may give further insight into the control of cell shape and how growth is directed either isotropically or polarly. The dimorphic fungi are defined as those organisms in which vegetative growth can occur in either a hyphal or budding mode depending on the environmental conditions [20]. The list of environmental effectors is rather exhaustive, the effect is often strain specific, and studies dealing with this aspect of dimorphism are numerous in the literature [101–105]. The following review section will consider the regulation of the physiological changes associated with the dimorphic transition and strategies that may be employed to control morphology.

Morphogenetic switching is not a unique feature of dimorphic organisms. During the process of germination, for example, the spores of many fungal species undergo a morphogenetic switch from isotropic expansion (during spore swelling) to polarized apical growth (when the germ tube is formed), whereafter, cell-surface expansion is confined to the hyphal tip [39]. In yeasts, alternating periods of polarly directed and isotropic growth are observed, as cells expand and then form buds [106]. *S. cerevisiae* can also form pseudohyphae under starvation conditions, especially nitrogen starvation [107]. Understanding how growth is directed polarly or isotropically may be the key to determining how the morphogenetic switch is controlled. Furthermore, identification of the genes involved in the control and regulation of dimorphism may point to targets for morphological engineering in industrially relevant filamentous fungi.

4.2.1
Biochemical Changes Associated with Dimorphism

Table 3 gives an overview of the biochemical changes associated with the dimorphic shift (yeast to mycelium transition) for a number of dimorphic organisms. Several of the changes may be considered to be common features of dimorphism, particularly alterations to the levels of molecules involved in signaling pathways and components, such as cAMP, which is known to operate as a secondary messenger. It has become apparent in yeasts that complex regulatory networks function to coordinate polarized morphogenesis with both the nuclear division cycle and cellular growth [108]. Signaling pathways involved in coordinating these events warrant further investigation, particularly the role of protein kinases which have been shown to have important roles in morphogenesis in filamentous fungi (discussed above).

In *Candida albicans* the expression of phospholipases increases when germ tubes are formed. This may be due to a role in tissue invasion, or, as has been observed in other organisms, phospholipases may participate in enzymatic cascades that generate highly active lipids used to transduce signals [109]. Manavanthu et al. [110], investigated the intracellular level of glutathione (which helps maintain the oxidation-reduction potential of the cell) during the yeast to mycelial conversion in this organism. While levels decreased significantly during the conversion, they concluded that this was not mediated by the inhibition of glutathione metabolic enzymes. Rather, this study indicated that the redox potential of the cell may regulate the activity of a key component(s) involved in the dimorphic conversion.

Of the regulatory components investigated, perhaps of most interest is cAMP, which has been studied in detail in *Mucor* species. cAMP is a small regulatory molecule, endogenously made within all cells [111], and is known to act primarily as an effector of protein kinases in eukaryotic cells [112]. In *Mucor* spp., yeast cells grown anaerobically contain high levels of cAMP compared to aerobically grown hyphae [103] and in *U. maydis*, the dimorphic transition has been shown to be regulated in part by cAMP-dependent protein kinase (protein

Table 3. Cellular biochemical changes associated with dimorphism in dimorphic fungi

Component	Organism	Effect	Reference
Phospholipase D	*C. albicans*	Dose-dependent stimulation of germ tube formation	119
Phospholipase B	*C. albicans*	Expression of PLB1 regulated as a function of morphogenetic transition	109
cAMP	*M. racemosus* and *M. rouxii*	CAMP levels four-fold higher in anaerobically grown yeasts than aerobically grown hyphae	120, 121
Glutamate dehydrogenase	*M. racemosus*	Only NAD (and not NADP) dependent GDH enhanced after induction of hyphal development	122
Ornithine decarboxylase and polyamine synthesis	*M. racemosus*	Initial 30–50-fold increase in activity of ODC throughout yeast to hyphal transition	123
	C. albicans	Higher polyamine levels required for hyphal growth	124
	U. maydis	Higher polyamine levels required for hyphal growth	125
Glutathione	*C. albicans*	Intracellular level decreased significantly during yeast-to-mycelial conversion	110
Cell wall mannans	*Mucor* spp.	Marked decrease in cell wall mannans as mycelial growth proceeds	106, 126, 127
Fatty acids and sterols	*M. genevensis and M. rouxii*	Hyphal cells have higher proportion of fatty acids and sterols than yeast cells	128, 129
Chitin	*C. albicans*	3–5 fold increase in cell wall chitin levels immediately following germ tube formation	130
Chitin synthase	*C. albicans*	Specific chitin synthase activity of hyphae estimated to be twice that of yeast cells	131
	M. racemosus	Rate of chitin and chitosan synthesis accelerated in mycelial cells	132
	M. circinelloides	Accumulation of Mcchs1 transcript during exponentially growing hyphal stage (not detected in yeast form)	133

kinase A) [113]. Additionally, mutations in genes encoding components of the cAMP pathway have been shown to confer dramatic morphological phenotypes in this organism [114]. External signals induce the switch from the yeast to hyphal growth form of *C. albicans*, and protein kinase A (PKA) has been shown to be required for the internal signaling leading to hyphal differentiation [115]. A similar mechanism has been found in *Mucor rouxii* [116]. The role of mitogen activated protein kinases (MAP kinases) in dimorphism has also been investigated, and recent studies describe similarities between *C. albicans* and *S. cere-*

visiae signal transduction pathways [117, 118]. In *S. cerevisiae*, two major signal transduction pathways (MAP kinase and cAMP regulated pathways) have been shown to be critical for differentiation of yeast cells to pseudohyphae.

Clearly, signaling and amplification of signals in response to external stimuli is a key feature of the control of morphogenetic switching in dimorphic organisms. The signaling pathways identified in dimorphic organisms show similarities to MAP kinase signaling pathways in other fungi; the role of these in morphogenesis of filamentous fungi has been previously discussed (Sect. 3.1.2). The evidence accumulated so far indicates that manipulating morphology is possible in yeasts and dimorphic fungi via metabolic engineering of signal transduction pathways (the molecular basis of this is discussed in Sect. 4.2.3). What remains to be seen is whether similar morphological engineering by this route is possible in filamentous fungi.

4.2.2 *Structural Changes Associated with Dimorphism*

As can be seen from Table 3, the composition of cytoskeleton and cell wall components has been observed to change during the dimorphic transition. In all dimorphic organisms, differences in the levels of cell wall components have been reported for yeast and mycelial forms of the organisms, the quantities and degree of change being strain dependent. Little is known about how the change is mediated for many of the components; however, there has been increasing interest in chitin and the activity of chitin synthases. As has been observed with filamentous fungi, this class of enzymes has an important role in determining hyphal morphology.

The chitin synthase enzymes of *C. albicans* are of particular interest in studies of dimorphism (and, thus, pathogenicity) in this organism. An increase in cell wall chitin levels (of the order of three- to fivefold) has been observed immediately following germ tube formation [130] and the specific chitin synthase activity of hyphae is estimated to be twice that of yeast cells [131]. Inhibition of key chitin synthases may, therefore, prevent the transition to the invasive hyphal form. Three chitin synthases (CHS1, CHS2, and CHS3) have been identified in *C. albicans*, with each isoenzyme performing a separate role at a distinct stage of the cell cycle. Chitin synthase gene expression is regulated differentially during yeast/hyphal transitions [131].

Munro et al. [131] investigated the expression of the *C. albicans* chitin synthase genes under conditions promoting both yeast and hyphal phases of growth. CHS1 was found to be expressed constitutively, at low levels in yeast and hyphal phases of growth while expression of CHS2 and CHS3 increased transiently during hyphal formation. However, Δ*chs2* and Δ*chs3* null mutants formed hyphae efficiently, indicating that these genes are not essential for hyphal formation. In addition, in wild type cells the chitin content of hyphae was maintained even when mRNA levels declined. Indeed, no clear relationship between up-regulated chitin synthase gene expression and changes in the chitin synthase activity, chitin content, or cell shape was found. In *U. maydis*, six chitin synthase genes have been identified [134]. Of these, *chs1-5* have been detected in both the yeast and mycelial forms and gene disruption has shown that each chitin synthase was non-essential for viability and had little effect on morphology. It has been suggested that the eucaryotic cytoskeletal elements (mainly actin and tubulin) may also play a role in *Mucor* dimorphism [103]. The effects on the levels of chitin and chitin synthases are summarized in Table 3. From the above discussion, it is concluded that in *C. albicans*, *U. maydis*, and *Mucor circinelloides* the dimorphic switch has not been attributed to a change in the level of expression of one chitin synthase in particular. This is likely to be due to the fact that compensation for loss of function can occur, as is the case with *Aspergilli* (Sect. 3.1.1), evidenced by the lack of mutant phenotype with *U. maydis* chitin synthase disrupted strains.

Changes in actin localization have also been observed to accompany hyphal development in *M. rouxii* [135], with a switch to polarized accumulation as germ tubes are formed [136]. Microtubules have not been visualized in electron micrographs of *Mucor* [103], although these have been studied in other dimorphic organisms. Cytoskeleton inhibitors were used to study the role of these components in morphogenesis of *C. albicans* [137] with apical cell elongation being arrested in the presence of microtubule and microfilament inhibitors. Results of this study showed, that it was microfilaments rather than microtubules that were essential for the cell elongation process. Microtubules are necessary, however, for the correct distribution of actin [138] and for the polarized localization of organelles [139], both processes having significant importance in morphogenesis [138] (Fig. 1).

4.2.3
Molecular Level Control of Dimorphism

Studies of the physiology of fungal dimorphism have identified the biochemical and structural changes associated with morphogenesis and have provided insight into how such differentiation may be controlled and regulated. An increasing number of studies employing molecular biology techniques have led to the identification of genes that are involved in control of dimorphism. In the case of *C. albicans* and *U. maydis*, both pathogenic organisms, this research has been fueled by the fact that the filamentous form is invasive and is, therefore, associated with pathogenicity. Additionally, a large body of work on the pseudohyphal growth of *S. cerevisiae* has contributed to our understanding of the general regulatory pathways involved in dimorphism. Some of the genes of interest and their roles are given in Table 4. In many cases, similarity with genes in filamentous fungi has been found. Clearly, analysis of the control of dimorphism at the molecular level is significant in contributing to our understanding of the processes involved in morphogenesis in filamentous fungi.

A number of the genes identified as having a role in dimorphic switching encode proteins involved in signaling pathways, such as TPK2 of *C. albicans* which encodes a catalytic subunit of PKA. Deletion of this gene blocks morphogenesis, whereas overexpression induces hyphal formation [115]. The *C. albicans mkc1* gene, also encoding a MAP kinase, has an additional role in biogenesis of the cell wall [118], with deletion resulting in cell wall defects. However, the cascades in-

Table 4.

Organism	Gene	Reference
Y. lipolytica	HOY1 required for hypha formation	140
C. albicans	TUP1 deletion results in a gene dosage dependent filamentous growth	141
C. albicans	MKC1	118
C. albicans and S. cerevisiae	INT1 disruption supresses hyphal growth	142
C. albicans	PHR1 required for morphogenesis by defining novel function for apical growth	143
C. albicans	Efg1p	144
C. albicans	CaRSR1 required for polar bud site selection in yeast form, germ tube emergence during mycelial transition and hyphal elongation.	145
C. albicans	TPK2 deletion blocks morphogenesis	115
C. albicans	MAP kinase gene MKC1	118

volved in the signaling pathways leading to polar rather than isotropic expansion have yet to be elucidated.

It must surely be of interest to investigate the similarity between genes that are expressed during the hyphal growth phase of dimorphic organisms and those of filamentous fungi. Several genes have been identified in dimorphic organisms that are necessary for hyphal or pseudohyphal growth, while in the filamentous fungi some genes (and their products) have been identified which are necessary for apical extension. Combination of the knowledge obtained for these two groups of organisms would greatly enhance our understanding of the factors involved in the control of morphology and provide a strong platform from which to continue with the metabolic engineering of morphology.

References

1. Adams HL, Thomas CR (1988) Biotechnol Bioeng 32:707
2. Packer HL, Thomas CR (1990) Biotechnol Bioeng 35:870
3. Thomas CR (1992) TIBTECH 10:343
4. Cox PW, Thomas CR (1992) Biotechnol Bioeng 32:945
5. Reichl U, King R, Gilles ED (1992) Biotechnol Bioeng 39:164
6. Paul GC, Thomas CR (1998) Adv Bioechem Eng/Biotechnol 60:1
7. Agger T, Spohr AB, Carlsen M, Nielsen J (1998) Biotechnol Bioeng 57:321
8. Krabben P, Nielsen J (1998) Adv Biochem Eng/Biotechnol 60:125
9. McIntyre M, Berry DR, McNeil B (1999) Enzyme Microbial Technol 25:447
10. Cui YQ, van der Lans RGJM, Luyben KCAM (1998) Biotechnol Bioeng 57:409
11. Paul GC, Kent CA, Thomas CR (1994) Trans I Chem E 72:95
12. Müller C, Spohr AB, Nielsen J (2000) Biotechnol Bioeng 67:390
13. Cox PW, Thomas CR (1999) Mycol Res 103:757
14. Bachewich C, Heath IB (1998) J Cell Sci 111:2005

15. Wolkow TD, Harris SD, Hamer JE (1996) J Cell Sci 109:2179
16. Christiansen T, Spohr AB, Nielsen J (1999) Biotechnol Bioeng 63:147
17. Spohr A, Dam-Mikkelsen C, Carlsen M, Nielsen J, Villadsen J (1998) Biotechnol Bioeng 58:541
18. Nielsen J, Villadsen J (1994) Bioreaction engineering principles. Plenum Press, New York
19. Heath IB (1995) Can J Bot 73:S131
20. Gow NAR (1995) Tip growth and polarity. In: Gow NAR, Gadd GM (eds) The growing fungus. Chapman and Hall, London, p 277
21. Wessels JGH (1993) Exp Phytol 123:397
22. Sietsma JH, Wessels JGH (1991) Cell wall assembly in fungal morphogenesis. In: Latgé JP, Boucias D (eds) Fungal cell wall and immune response. Springer, Berlin Heidelberg New York, p 81
23. McKerracher LJ, Heath IB (1987) Exp Mycol 11:79
24. Jackson SL, Heath IB (1992) Protoplasma 170:46
25. Heath IB (1990) Int Rev Cytol 123:95
26. Riquelme M, Reynaga-Pena CG, Gierz G, Bartnicki-Garcia S (1999) Fungal Genet Biol 24:101
27. Roberson RW, Fuller MS (1988) Protoplasma 146:143
28. Howard RJ, Aist JR (1979) J Ultrastr Res 66:224
29. Grove SN, Bracker CE (1970) J Bacteriol 104:989
30. Parton RM, Fischer S, Malhó R, Papasouliotis O, Jelitto TC, Leonard T, Read ND (1997) J Cell Sci 110:1187
31. Jackson SL, Heath IB (1993) Protoplasma 173:23
32. Robson GD, Wiebe MG, Trinci APJ (1991) J Gen Microbiol 137:963
33. Borgia PT, Dodge CL (1992) J Bacteriol 174:377
34. Gooday GW (1971) J Gen Microbiol 67:125
35. Bartnicki-Garcia S, Lippman E (1969) Science 165:302
36. Trinci APJ (1978) The duplication cycle and vegetative development in moulds. In: Smith JE, Berry DR (eds) The filamentous fungi, vol 3. Arnold, London, p 132
37. Kaminskyj SGW, Hamer JE (1998) Genetics 148:669
38. Wessels JGH (1990) Role of cell wall architecture in fungal tip growth generation. In: Heath IB (ed) Tip growth in plant and fungal cells. Academic Press, New York, p 1
39. Harris SD, Hofmann AF, Tedford HW, Maurice PL (1999) Genetics 151:1015
40. Mata J, Nurse P (1998) Trends Cell Biol 8:163
41. Schmidt A, Hall MN (1998) Ann Rev Cell Dev Biol 14:305
42. Torralba S, Raudaskoski M, Pedregosa AM, Laborda F (1998) Microbiology 144:45
43. Reissig JL, Kinney SG (1983) J Bacteriol 154:1397
44. Rasmussen H, Goodman DBP (1977) Physiol Rev 57:421
45. Katz D, Goldstein D, Rosenberger RF (1972) J Bacteriol 109:1097
46. Suelmann R, Sievers N, Fischer R (1997) Mol Microbiol 25:757
47. Xiang X, Beckwith SM, Morris R (1994) Proc Natl Acad Sci USA 91:2100
48. Xiang X, Morris R (1999) Curr Opinion Microbiol 2:636
49. Robb MJ, Wilson MA, Vierula PJ (1995) Mol Gen Genet 24:583
50. Nielsen J (1995) Physiological engineering aspects of *Penicillium chrysogenum*. DSc Thesis, Technical University of Denmark
51. Nielsen J, Carlsen M (1996) Fungal pellets. In: Willaert RG, Baron GV, De Backer L (eds) Immobilised living cell systems: modelling and experimental methods. Wiley, New York, p 273
52. Spohr A, Agger T, Carlsen M, Nielsen J (1998) Quantitative morphology of filamentous organisms. In: Wilkinson MHF, Schut F (eds) Digital image analysis of microbes: imaging, morphometry, fluorometry and motility techniques and applications. Wiley, New York, p 373
53. Carlsen M, Spohr AB, Nielsen J, Villadsen J (1996) Biotechnol Bioeng 49:266
54. Paul GC, Kent CA, Thomas CR (1994) Trans I Chem E 72C:95
55. Tucker KG, Thomas CR (1993) Trans I Chem E 171:111

56. Braun S, Vecht-Lifshitz SE (1991) TIBTECH 9:63
57. Shamlou PA, Makagiansar HY, Ison AP, Lilly MD, Thomas CR (1994) Chem Eng Sci 49:2621
58. Makagiansar HY, Shamlou PA, Thomas CR, Lilly MD (1993) Bioprocess Eng 9:83
59. Smith JJ, Lilly MD, Fox RI (1990) Biotechnol Bioeng 35:1011
60. Paul GC, Kent CA, Thomas CR (1992) Trans I Chem E 70:13
61. Vanhoutte B, Pons MN, Thomas CR, Louvel L, Vivier H (1995) Biotechnol Bioeng 48:1
62. Sietsma JH, Wösten HAB, Wessels JGH (1995) Can J Bot 73:S388
63. Peberdy JF (1994) TIBTECH 12:50
64. Wösten HAB, Moukha SM, Sietsma JH, Wessels JGH (1991) J Gen Microbiol 137:2017
65. Wongwicharn A, McNeil B, Harvey LM (1999) Biotechnol Bioeng 65:416
66. Johansen CL, Coolen L, Hunik JH (1998) Biotechnol Prog 14:233
67. Bocking SP, Wiebe MG, Robson GD, Hansen K, Christiansen LH, Trinci APJ (1999) Biotechnol Bioeng 65:638
68. Spohr A, Carlsen M, Nielsen J, Villadsen J (1997) Biotechnol Letts 19:257
69. Fiddy C, Trinci APJ (1971) J Gen Microbiol 97:169
70. Amanullah A, Blair R, Nienow AW, Thomas CR (1999) Biotechnol Bioeng 62:434
71. Mitard A, Riba JP (1988) Biotechnol Bioeng 32:835
72. Metz B, Kossen NWF (1977) Biotechnol Bioeng 19:781
73. Riley GL, Tucker KG, Paul GC, Thomas CR (2000) Biotechnol Bioeng 68:160
74. Gordon CL, Khalaj V, Ram AFJ, Archer DB, Brookman JL, Trinci APJ, Jeenes DJ, Doonan JH, Wells B, Punt PJ, van den Hondel CAMJJ, Robson G (2000) Microbiology 146:415
75. Henriksen AS, Even S, Müller C, Punt PJ, van den Hondel CAMJJ, Nielsen J (1999) Microbiology 145:729
76. Harris SD, Hamer L, Sharpless KE, Hamer JE (1997) EMBO J 16:3474
77. Momany M, Westfall PJ, Abramowsky G (1999) Genetics 151:557
78. Stark MJ (1996) Yeast 12:1647
79. Dickman MB, Yarden O (1999) Fungal Genet Biol 26:99
80. Madhani HD, Fink GR (1998) Trends Cell Biol 8:348
81. Bruno KS, Aramayo R, Minke PF, Metzenberg RL, Plamann M (1996) EMBO J 15:5772
82. Lee IH, Walline RG, Plamann M (1998) Mol Microbiol 29:209
83. Rudall KM (1969) J Polymer Science 28X:83
84. Kang MS, Elango N, Mattia E, Au-Young J, Robbins PW, Cabib E (1984) J Biol Chem 259:14,966
85. Burnett JH (1979) Aspects of the structure and growth of hyphal walls. In: Burnett JH, Trinci APJ (eds) Fungal walls and hyphal growth. Cambridge University Press, Cambridge, p 1
86. Fujiwara M, Horiuchi H, Ohta A, Takagi M (1997) Biochem Biophys Res Commun, 236:75
87. Bowen AR, Chen-Wu JL, Momany M, Young R, Szaniszlo PJ, Robbins PW (1992) Proc Natl Acad Sci USA 89:519
88. Horiuchi H, Takagi M (1999) Chitin synthase genes of Aspergillus species. In: Brakhage AA, Jahn B, Schmidt A (eds) *Aspergillus fumigatus*: Contrib Microbiol. Basel Karger, Basel, p 193
89. Borgia PT, Iartchouk N, Riggle PJ, Winter KR, Koltin Y, Bulawa CE (1996) Fungal Genet Biol 20:193
90. Horiuchi H, Fujiwara M, Yamashita S, Otha A, Takagi M (1999) J Bacteriol 181:3721
91. De Marini DJ, Adams AEM, Fares H, De Virgilio C, Valle G, Vhuang JS, Pringle JR (1997) J Cell Biol 139:75
92. Choi W-J, Santos B, Duran A, Cabib E (1994) Mol Cell Biol 14:7685
93. Mellado E, Aufauvre-Brown A, Specht CA, Robbins PW, Holden DW (1996) Mol Gen Genet 246:353
94. Iten W, Matile P (1970) J Gen Microbiol 62:203
95. Horsch M, Rast DM (1993) Allosteric activation of chitin synthase by *N*-acetylglucosamine: a mechanistic study. In: Muzzarilli RAA (ed) Chitin enzymology. European Chitin Society, Ancona, Italy, p 47

96. Specht CA, Liu Y, Robbins PW, Bulawa CE, Iartchouk N, Winter KA, Riggle PJ, Rhodes JC, Dodge CL, Culp DW, Borgia PT (1996) Fungal Genet Biol 20:153
97. Culp DW, Dodge CL, Yihong M, Li L, Demet S-O, Borgia PT (2000) FEMS Microbiol Lett 182:349
98. Motoyama T, Sudoh M, Horiuchi H, Ohta A, Takagi M (1994) Biosci Biotech Biochem 58:1685
99. Motoyama T, Fujiwara M, Kojima N, Horiuchi H, Ohta A, Takagi M (1997) Mol Gen Genet 253:520
100. Yanai K, Kojima N, Takaya N, Horiuchi H, Ohta A, Takagi M (1994) Biosci Biotech Biochem 58:1828
101. Northrop FD, Ljubojevic S, Davies JM (1997) Microbiology 143:3757
102. Gow NAR (1997) Curr Top Med Mycol 8:43
103. Orlowski M (1991) Microbiol Rev 55:234
104. Odds FC (1985) Crit Rev Microbiol 12:45
105. Bartnicki-Garcia S (1968) J Bacteriol 96:1586
106. Madden K, Snyder M (1998) Ann Rev Microbiol 52:687
107. Brown CM, Hough JS (1965) Nature 206:676
108. Kron SJ, Gow NAR (1995) Curr Opin Cell Biol 7:845
109. Hoover CI, Jantaopour MJ, Newport G, Agabian N, Fisher SJ (1998) FEMS Microbiol Lett 167:163
110. Manavanthu E, Duncan C, Porte Q, Gunasekaran M (1996) Mycopathologia 135:79
111. Rickenberg HV(1974) Annu Rev Microbiol 28:353
112. Robinson GA, Butcher RW, Sutherland EW (1971) Cyclic AMP. Academic Press, New York
113. Dürrenberger F, Kronstad J (1999) Mol Gen Genet 261:281
114. Dürrenberger F, Wong K, Kronstad J (1998) Proc Natl Acad Sci USA 95:5684
115. Sonneborn A, Bockmuhl DP, Gerads M, Kurpanek K, Sanglard D, Ernst JF (2000) Mol Microbiol 35:386
116. Rossi S, Moreno S (1994) Eur J Biochem 222:501
117. Alonso-Monge R, Navarro-García F, Molero G, Diez-Orejas R, Gustin M, Pla J, Sánchez M, Nombela C (1999) J Bacteriol 181:3058
118. Navarro-García F, Alonso-Monge R, Rico H, Pla J, Sentandreu R, Nombela C (1998) Microbiology 144:411
119. McLain N, Dolan JW (1997) Microbiology 143:3521
120. Larsen AD, Sypherd PS (1974) J Bacteriol 117:432
121. Paveto C, Epstein A, Passeron S (1975) Mol Cell Biochem 65:55
122. Peters J, Sypherd PS (1979) J Bacteriol 137:1137
123. Inderlied CB, Cihlar RL, Sypherd PS (1980) J Bacteriol 141:699
124. Herrero AB, López MC, García S, Schmidt A, Spaltmann F, Ruiz-Herrera J, Dominguez A (1999) Infect Immun 67:4870
125. Guevara-Olvera L, Xonocostle-Cázares B, Ruiz-Herrera J (1997) Microbiology 143: 2237
126. Bartnicki-Garcia S, Nickerson WJ (1962) J Bacteriol 84:841
127. Sypherd PS, Borgia PT, Paznokas JL (1978) Adv Microbial Phys 18:67
128. Gordon PA, Stewart PR, Clark-Walker GD (1971) J Bacteriol 107:114
129. Safe S, Caldwell J (1975) Can J Microbiol 21:79
130. Chen-Wu JL, Zwicker J, Bowen AR, Robbins PW (1992) Mol Microbiol 6:497
131. Munro CA, Schofield DA, Gooday GW, Gow NAR (1998) Microbiology 144:391
132. Domek DB, Borgia PT (1981) J Bacteriol 146:945
133. Lopez-Matas MA, Eslava AP, Diaz-Minguez JM (2000) Curr Microbiol 40:169
134. Xonocostle-Cazares B, Specht CA, Robbins PW, Liu Y, León C, Ruiz-Herrera J (1997) Fungal Genet Biol 22:199
135. Sabanero M, Zazueta R (1989) FEMS Microbiol Lett 60:227
136. Hasek J, Bartnicki-Garcia S (1994) Arch Microbiol 161:363
137. Yokoyama K, Kaji H, Miyaji M (1990) J Gen Microbiol 136:1067
138. Akashi T, Kanve T, Tanaka K (1994) Microbiology 140:271

139. Steinberg G, Schliwa M, Lehmler C, Bölker M, Kahmann R, McIntosh JR (1998) J Cell Sci 111:2235
140. Torres-Guzman JC, Dominguez A (1997) Mol Cell Biol 17:6283
141. Braun BR, Johnson AD (1997) Science 277:105
142. Gale CA, Bendel CM, McClellan M, Hauser M, Becker JM, Berman J, Hostletter MK (1998) Science 27:1355
143. Saporito-Irwin SM, Birse CE, Sypherd PS, Fonzi WA (1995) Mol Cell Biol 15:601
144. Stoldt VR, Sonneborn A, Leuker CE, Ernst JF (1997) EMBO J 16:1982
145. Yaar L, Mevarech M, Koltin Y (1997) Microbiology 143:3033

Received: November 2000

Evolutionary Engineering of Industrially Important Microbial Phenotypes

Uwe Sauer

Institute of Biotechnology, ETH Zürich, 8093 Zürich, Switzerland,
e-mail: sauer@biotech.biol.ethz.ch

The tremendous complexity of dynamic interactions in cellular systems often impedes practical applications of metabolic engineering that are largely based on available molecular or functional knowledge. In contrast, evolutionary engineering follows nature's 'engineering' principle by variation and selection. Thus, it is a complementary strategy that offers compelling scientific and applied advantages for strain development and process optimization, provided a desired phenotype is amenable to direct or indirect selection. In addition to simple empirical strain development by random mutation and direct selection on plates, evolutionary engineering also encompasses recombination and continuous evolution of large populations over many generations. Two distinct evolutionary engineering applications are likely to gain more relevance in the future: first, as an integral component in metabolic engineering of strains with improved phenotypes, and second, to elucidate the molecular basis of desired phenotypes for subsequent transfer to other hosts. The latter will profit from the broader availability of recently developed methodologies for global response analysis at the genetic and metabolic level. These methodologies facilitate identification of the molecular basis of evolved phenotypes. It is anticipated that, together with novel analytical techniques, bioinformatics, and computer modeling of cellular functions and activities, evolutionary engineering is likely to find its place in the metabolic engineer's toolbox for research and strain development. This review presents evolutionary engineering of whole cells as an emerging methodology that draws on the latest advances from a wide range of scientific and technical disciplines.

Keywords. Adaptation, Directed evolution, Evolutionary engineering, Metabolic engineering, Selection

1 **Introduction** 130

2 **Mutagenesis and Recombination** 134

2.1 Physiologically Enhanced Spontaneous Mutagenesis 135
2.2 Chemical or Radiation Induced Mutagenesis 135
2.3 Mutator Strains 137
2.4 Tagged Mutagenesis 138
2.5 In Vivo Recombination 139

3 **Selection** 140

3.1 Natural Evolution 141
3.2 Solid Media 143
3.3 Batch 143
3.4 Microcolonization 144

Advances in Biochemical Engineering/
Biotechnology, Vol. 73
Managing Editor: Th. Scheper

3.5 Chemostat 145
3.6 Other Continuous Culture Devices 148
3.7 Fitness Landscapes and Effective Means of Conquering Fitness Peaks 149
3.8 Screening of Desired Variants from Evolved Populations 151

4 Evolutionary Engineering of Simple Cellular Subsystems 153

5 Evolutionary Engineering of Complex Cellular Subsystems 157

5.1 Resistance to Environmental Stress 157
5.2 Resistance to Metabolic Stress 158
5.3 Plasmid Stability 159
5.4 Mycelial Morphology 160
5.5 General Physiological Properties 161

6 Outlook 163

References 166

List of Abbreviations

BOICS	Brown and Oliver interactive chemostat selection
bp	base pair
DNA	desoxyribonucleic acid
EMS	ethyl methane sulfonate
IS	insertion element
kb	kilo base pairs
MS	mass spectrometry
NTG	nitroso-methyl guanidine
PCR	polymerase chain reaction
PTS	phosphotransferase system
mRNA	messenger ribonucleic acid
UV	ultra violet

1 Introduction

Research programs attempting to improve industrial properties of microorganisms were initially focused on strain selection after classical mutagenesis but the advent of recombinant DNA technology has dramatically expanded our capabilities and affected most contemporary research. In the area of cellular functions, rational applications of recombinant DNA technology are referred to today as *metabolic engineering* [1] and several successful approaches are reviewed in other contributions of this volume and elsewhere [1–3]. However, the complex nature of the highly interactive and elaborate informational and biochemical networks that govern cellular function presents major challenges to any metabolic engineering attempt and, in fact, has hampered successful industrial implementation in many cases. Although algorithms and modeling frameworks are being developed to improve identification of effective genetic changes, the extensive molecular and mechanistic information that is required to guide *constructive* metabolic engineering approaches remains a main drawback to rational, deductive strategies. An additional problem arises from the difficulty of predicting secondary responses or side-effects due to lack of knowledge of inter-related regulatory and metabolic processes in a cell. Experimental experience in both academic and industrial labs has shown that secondary responses to genetic modifications often occur in pathways or reactions that are seemingly unrelated to the target, thereby confounding the rational strategies [1, 4, 5].

Very similar problems were associated with rational protein engineering, and so it is both stimulating and instructive to consider recent developments in this related field. Much like current constructive metabolic engineering, previous strategies in protein engineering mainly attempted a rational design via defined, site-directed changes based on structural and mechanistic information [6]. Because such fundamental information is often not available, commercial applications were limited. Moreover, many rational attempts to alter protein properties failed because either the chosen target amino acids were not appropriate or the introduced substitutions exerted unanticipated influences on structure or function. Today, novel high-throughput techniques and discovery approaches including biodiversity screening, genomic sequencing, phage display, in vitro screening methods, and directed evolution are rapidly replacing or complementing rational design in industrial biocatalysis [7, 8].

One of the most promising strategies in protein engineering is directed evolution, which has been successfully employed to improve existing protein functions several thousand-fold and also to tailor completely new, artificial enzyme properties (but, so far, not de novo functions) that are not found in the natural environment [9, 10]. Such capabilities are also useful for metabolic engineering. Directed evolution is generally understood as the use of repeated cycles of creating genetic diversity and sifting pools of variants by immediate selection or screening to recover only those with a desired functional property (Fig. 1). For a general introduction to the field see [11]. A major technological advance in evolutionary protein engineering was the introduction of in vitro recombination by 'hybrid PCR', for example by DNA shuffling, because multiple, related starting points can be used rather than a single gene [9]. The power of recombination arises from the possibility of removing neutral or deleterious mutations as well as preserving useful mutations, which may improve the desired property in a synergistic fashion when combined. The generated libraries of chimeric genes are searched either by selection, in which a protein is linked to host survival, or, if that is not feasible, by direct screening, which is basically selection at the single variant level [12]. This evolutionary concept has already been extended from single proteins to entire pathways [11] and the next frontiers are the shuffling of entire viral or even microbial genomes and directed evolution of novel pathways [13, 14].

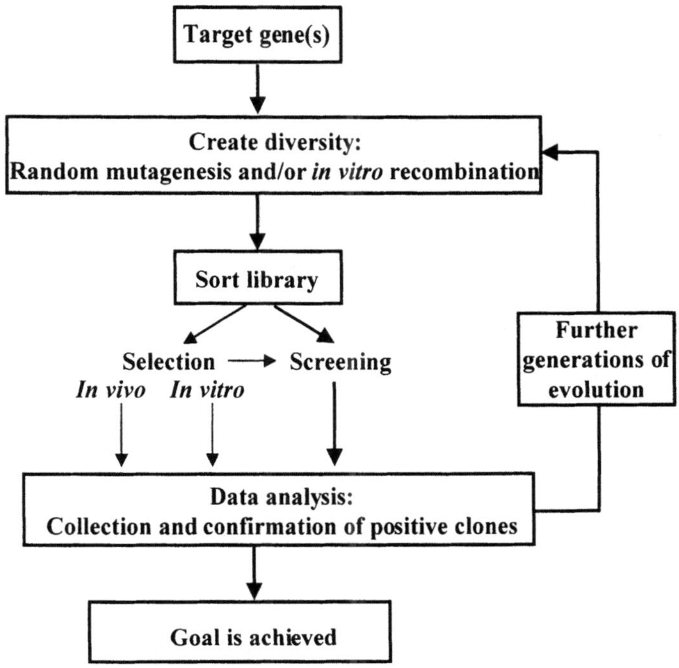

Fig. 1. Flow chart for directed enzyme evolution. Reproduced with permission from Zhao et al. [146]

Obviously, engineering of proteins shares many features with engineering of whole cells and so it is quite instructive to consider the suitability of evolutionary methods for metabolic engineering. In discussing evolutionary approaches it is helpful to employ the concept of *fitness landscapes* [15–17], which are topological representations of biological fitness in a given environment. Each genotype (or protein sequence) is associated with a fitness value (the phenotype) and the distribution of these functional values over the sequence space of all genotypes constitutes a fitness landscape. In natural evolution, fitness applies principally to the reproductive success of a species, and thus is rarely assigned to single genes. When referring to well-defined, *desired* characteristics of proteins or cells, the term *local* fitness landscape is frequently used to indicate that a particular fitness landscape is projected onto the sequence space. Thus, fitness is generally used in a much more restricted sense in applied evolutionary approaches. As a practical matter, sequence spaces are extraordinarily large, because the number of all possible sequences N is an exponential function of the number of information units λ (i.e., 4 nucleotides for DNA and 20 amino acids for proteins) and the length of the sequence (ν), according to

$$N = \lambda^{\nu} \tag{1}$$

Thus, even a single protein with 230 amino acids spans a sequence space of 10^{300} points [8, 18], which is not fully accessible by any experimental method. Cells are several order of magnitude more complex than proteins, and so the sequence spaces of even very modest genetic changes are dauntingly large. Fortunately, evolution proceeds not by exploring all possible variants but by incorporating single mutations, selecting the fittest of those, and then expanding the population and incorporating additional alterations [15,19]. Therefore, most applied evolutionary strategies assume the existence of an evolutionary path that yields detectably improved fitness for each mutation that is required for a desired phenotypic change. Thus, it resembles natural evolution which is, in effect, a method of searching among an enormous number of possibilities for small, step-wise improvements that allow organisms to survive better and reproduce in their environments.

The basic concept of directed evolution is also evident in classical, empirical strain development by classical, random mutagenesis and direct selection on plates. This approach has a long history of success in industrial strain development, in particular in the absence of extensive genetic or physiological information. The best example of this is probably the greater than 4000-fold improvement of penicillin titers via empirical strain improvement [20–22]. Empirical procedures are particularly well suited for relieving feedback inhibition in biosynthetic pathways because simple and direct selection schemes can be applied, for instance resistance to toxic analogs of metabolic intermediates (antimetabolites). Unfortunately, most desired phenotypes cannot be selected by simply increasing resistance towards a challenging agent. Analytical screening for desired phenotypes in random variants is not an alternative, because it does not provide access to any significant fraction of most local cellular fitness landscapes. Another disadvantage of extensive passage through cycles of mutagenesis and selection is the concomitant accumulation of unfavorable mutations, which eventually leads to highly specialized but crippled strains, a commonly observed phenomenon. This cost to asexual evolution of small populations is known as Müller's ratchet [23], the underlying principle for reductive evolution of resident genomes such as endosymbionts or cellular organelles [24].

These problems of step-wise directed evolution with whole cells can potentially be solved by two strategies that are also at work in nature: recombination and *continuous* selection in large populations for many generations. In the first strategy, recombination of genetic elements and subsequent selection is used to combine beneficial mutations from different variants in one strain and to reduce the mutational load by eliminating deleterious mutations, thereby potentially avoiding Müller's ratchet. Consequently, additional beneficial mutations need not be 'rediscovered' in a selected strain to become incorporated in future generations. The most powerful tool to navigate fitness landscapes in protein engineering, in vitro recombination [18], is presently restricted to subgenomic elements that can be amplified by PCR, and thus is not applicable to entire microbial genomes. Although microorganisms are naturally capable of in vivo recombination, this process has rarely been exploited for directed evolution of biotechnologically relevant phenotypes.

In the second strategy, continuous in vivo evolution of entire populations circumvents passage through the single variant level after each mutation-selection cycle. This is possible because microorganisms are self-replicating, unlike proteins, so that the phenotype is coupled to the genotype (at least as a first approximation). Due to their small size, microbial laboratory populations are large, exceeding 10^{11} individuals per liter (solutions with less than 5×10^9 cells per liter appear completely clear to the human eye), so that continuous evolution can be far more effective than step-wise procedures. The steady interplay between selection by the artificially posed conditions and mixed populations of continuously occurring genetic variants gives such continuous evolution its direction – potentially towards a desired phenotype, provided a pertinent selection scheme can be devised.

Due to the immense size of sequence spaces, evolutionary paths to improved variants may go astray or reach suboptimal solutions. This is intuitively recognized, since most evolutionary strategies are initiated with a phenotype that is already close to the desired one and thus may be considered more as engineering than as design strategies. Unlike step-wise evolutionary protein engineering, successful evolution of improved cells cannot be expected to lead to fully developed processes or products, but rather to constitute an important intermediate step in an engineering strategy. In industrial practice, strain developmental problems are often solved by synergistic application of metabolic engineering and empirical mutagenesis/selection. Thus, it can be anticipated that even more elaborate evolutionary methods will likewise be most powerful if used in combination with, or as the basis for, metabolic engineering to create synergistic effects for process improvement. I will refer to such applications of evolutionary techniques to microbial properties in a biotechnological context as *evolutionary engineering*, a term introduced by Butler et al. [25]. A prerequisite for any such evolutionary engineering is a selection scheme that directly or indirectly favors a desired phenotype.

A comprehensive understanding of microbial evolution combined with the ability to apply its principles to experimental systems are prerequisites to creating or optimizing microbial phenotypes with scientific or applied value by evolutionary engineering. Thus, without attempting to review comprehensively the literature on microbial evolution, this review highlights key concepts in designing and running evolutionary engineering programs. Furthermore, recent studies that employ evolutionary strategies to generate desired, heritable microbial phenotypes are reviewed and discussed. Applications of empirical mutagenesis/screening were recently reviewed [20–22], and so are not covered here. Finally, novel analytical procedures that may facilitate identification of the molecular basis of evolved phenotypes and thus impact evolutionary engineering will be briefly discussed.

2
Mutagenesis and Recombination

Mutations are a double-edged sword – the ultimate source of all genetic variation upon which any evolutionary process depends, yet the vast majority either have no apparent effect or are harmful, and so the rate of mutagenesis has to be appropriately tuned to design an efficient evolutionary process. Spontaneous mutations in microbial populations occur much less frequently than in viruses – generally at about 0.003 point mutations per genome (independent of its size) and round of replication [26]. Notable exceptions are the so-called hypermutable genes in pathogenic organisms that are prone to mutation through various specific mechanisms [27]. At first glance, accelerated generation of variation, or an increase in the population size for that matter, thus appears to be advantageous for practical application of continuous evolution. In asexual populations, however, higher mutation rates need not accelerate the pace of evolutionary adaptation [28], which is the underlying principle of selection for new or improved phenotypes. Examples are populations in which two different lineages of beneficial mutations interfere with one another's spread. Because the two mutations cannot be combined into the same lineage without recombination, such clonal interference imposes a speed limit on adaptive evolution. In small or initially well-adapted populations that spend long times waiting for beneficial mutations, on the other hand, an increase in the mutation rate may effectively accelerate the evolutionary process. Mutability is genetically determined like any other property, hence mutability itself can be affected by environmental (Sects. 2.1 and 2.2) or genetic (Sects. 2.3 and 2.4) manipulations, including recombination (Sect. 2.5).

2.1 Physiologically Enhanced Spontaneous Mutagenesis

Spontaneous alterations in the inheritable genetic sequence may result from a multitude of causes and mechanisms that can be grouped into three categories – (i) small local changes, (ii) DNA rearrangements, and (iii) horizontal DNA transfer, as illustrated in Table 1 [29, 30]. While the overall rate of spontaneous mutagenesis is usually rather stable and low [26], it may rise considerably under certain circumstances and modulation of environmental conditions provides a convenient means to accelerate this rate. For example, the global rate of mutagenesis in a population increases during adverse environmental conditions, for instance metabolic stress or stationary phase [29, 31]. Such environmental stimuli induce enzyme systems, mostly DNA polymerases that are designed to generate mutations, such as the SOS DNA repair system. Unlike the replicative DNA polymerases, which faithfully copy DNA sequences, these polymerases introduce errors at high rates, thereby increasing the genetic diversity and adaptation potential of the endangered population. Less well recognized is the fact that glucose repression may also reduce spontaneous mutagenesis, as the rate at which spontaneous *E. coli* mutants occur is several-fold lower on glucose than, for example, on glycerol [32]. While such environmental factors can accelerate the rate of mutagenesis, they will inevitably also influence the process of selection.

2.2 Chemical or Radiation Induced Mutagenesis

Induction of mutagenesis by chemicals or radiation treatment is frequently used because it is technically simple and widely applicable to almost any organism [29]. Most chemical mutagens preferentially introduce certain types of mutations such as exchange of specific nucleotides or frame-shifts, but many, including ethyl methane sulfonate (EMS), can also induce deletions of considerable

Table 1. Classification of mutations, their origins, and potential effects

Type of change	Length	Source of mutation	Effects[a]
Small local changes			
Substitution Insertion Deletion Duplication	1 bp 1 to several bp 1 to several bp 1 to several bp	Spontaneous mutagenesis Replication infidelities	Gene silencing Gene expression Cryptic gene activation Altered protein specificities
DNA rearrangements			
Inversion Duplication Insertion Deletion Excision	Several bp up to several kb	Homologous recombination Mobile genetic elements (i. e. IS elements, transposons)	Gene silencing Gene expression Cryptic gene activation Gene dosage Gene organization Gene mobilization Domain fusion Domain swapping
DNA acquisition			
Horizontal DNA transfer	Several kb up to hundreds of kb	Transformation Conjugation Transduction (phage-mediated)	Increase of total genetic information content Gene silencing

[a] A particular source of mutation is not necessarily capable of causing all listed effects.

length. For example, about 13% of the EMS-induced mutations in *Caenorhabditis elegans* are reported to be DNA rearrangements, and most of these are deletions with an average size of 1300 bp and a broad size range [33]. The use of nitroso-methyl guanidine (NTG), on the other hand, typically results in closely linked mutations in one clone due to its specificity for mutating DNA at the replication fork. Another factor that needs to be borne in mind is the phenomenon of biological mutagen specificity, whereby a given mutagenic treatment preferentially mutates certain parts of the genome [21]. Thus, for repetitive uses, it is advisable to change mutagens periodically, to take advantage of their presumably different mechanisms of action. The preferred mutagens for most applications are far UV, EMS, and NTG, because they induce a great variety of molecular alterations with no apparent specificity for genomic subregions [34].

For efficient evolutionary engineering, mutagenic treatment with an optimum dose of mutagen is particularly critical when performing successive rounds of mutagenesis and selection [34]. While the primary requirement is to increase the proportion of mutants in the surviving population, the optimum dose yields the highest proportion of desirable mutants. Although the optimum dose may be difficult to estimate for complex or difficult-to-detect phenotypes, related but easily scorable phenotypes may be used to help determine the optimum range. Any mutagenic treatment will give a dose response curve similar to

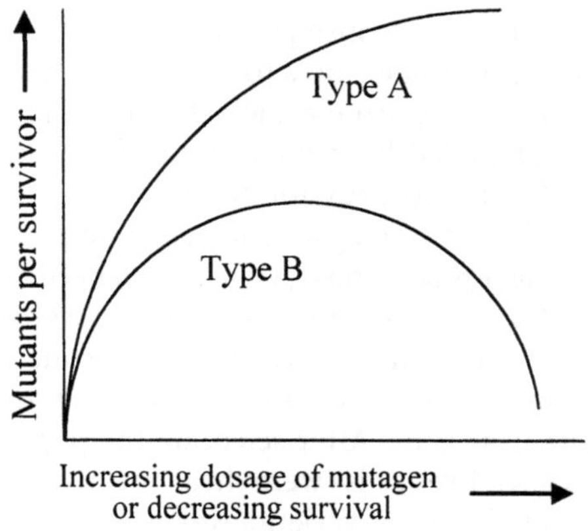

Fig. 2. Typical mutation kinetics curves. Reproduced from Rowlands [34]

either curve A or B in Fig. 2, wherein the type of curve appears to depend on the scored phenotype rather than on the mutagenic treatment used. While suboptimal mutagen doses will obviously create less diversity, overdoses of mutagens will simply kill the cells. Moreover, dosages even slightly above the optimum will increase the frequency at which neutral or potentially harmful mutations also become incorporated into the selected mutants. This is because advantageous adaptive mutations that occur in the background of neutral or weakly counter-selected mutations allow these undesired mutations to hitchhike along [35].

2.3 Mutator Strains

A fascinating option for accelerating continuous evolution is the use of so-called mutator strains, which are characterized by frequencies of spontaneous mutagenesis that are orders of magnitude higher than usual. In many cases, such mutations promote more rapid adaptive evolution, and mutator strains were shown to outcompete quickly the wild-type in glucose-limited environments [36]. In fact, mutations in mutator genes occur frequently in populations that are propagated over extended periods under identical conditions [37]. Intuitively, such mutations appear advantageous for evolutionary adaptation but their frequent occurrence in adapted populations is more likely circumstantial, resulting from numerous opportunities for the mutator mutation to hitchhike along with beneficial mutations to which they are genetically linked under these conditions [28]. Thus, mutators do not necessarily accelerate the pace of evolutionary adaptation, as was discussed more generally for spontaneous mutations before. Nevertheless, mutator genotypes can be very valuable in well-designed continuous evolution strategies, such as when evolving populations would be expected to spend most of their time waiting for beneficial mutations (e.g., [38]), as may be the case with already well-adapted strains.

A negative aspect of using such highly mutating strains is the potential accumulation of deleterious mutations that may reduce overall fitness [39] and their inherent phenotypic instability. Consequently, mutator genotypes have more frequently been used as convenient tools to introduce mutations into plasmid- or phage-encoded recombinant proteins, which can simply be separated from the background of accumulated harmful and neutral genomic mutations [40, 41]. A potentially very useful strategy for accelerated continuous evolution of particular genes is based upon propagating a phagemid population in a mutator strain. In one study using a β-lactamase, which confers resistance to the antibiotic cefotaxime, up to 1000-fold more resistant variants were obtained after a few weeks of selection in media with increasing cefotaxime concentration [42]. Briefly, a mutator strain was co-infected with a helper phage and a phagemid that carries the β-lactamase gene. After selecting the population for increased resistance to cefotaxime, live cells were heat-inactivated and the evolved phagemid population of about 10^6 variants was used to infect a fresh mutator host. This procedure ensured that only mutations within the phagemid genome are transferred into the next evolutionary cycle.

Many genes that cause a mutator phenotype are involved in repair or error avoidance systems, and bacterial mutator genes were recently reviewed by Miller [43]. For example, mutations in the *E. coli dnaQ* gene, which encodes the exonuclease activity-providing ε subunit of DNA polymerase III, impair the proofreading activity and hence lead to a very strong mutator phenotype. Similarly, mutations in components involved in the mismatch repair system also cause a strong mutator phenotype. Mutator genes in the eukaryote *S. cerevisiae* include the MMS2 gene (involved in postreplication repair) [44] and the POL30 gene, which is involved in mutation suppression [45]. The mutations caused by mutator phenotypes are mostly base transitions and frameshifts, but may also include deletions. At least for *E. coli*, such mutator strains can either be generated by defined genetic manipulations or by direct selection on a single plate [46].

2.4 Tagged Mutagenesis

All heretofore mentioned mutagenesis procedures have a serious disadvantage in that it is difficult to locate the modification, unless phenotypic characterization and a known gene-function relationship provide a clear lead. The use of tagged mutagenesis is one approach to facilitating the transfer of an evolved phenotype by metabolic engineering to others strains or organisms. For this purpose, a broad range of transposable elements is available, including genetically engineered mini-transposons [47]. These DNA elements catalyze their own movement, or transposition, to a location within a chromosome or, in certain cases, preferentially within extrachromosomal elements [48]. In addition to gene disruption, such transposable elements may also be used for random gene overexpression if equipped with suitable outward-oriented promoters. Most transposons, however, exhibit some degree of target preference and their capability for multiple insertions within one strain is usually limited.

An alternate strategy for mutagenesis and gene tagging is based on random insertion of unique, short DNA fragments ('signatures'), which is normally used for parallel identification of important, habitat-specific genes by negative selection [49]. Because insertional inactivation of genes may also improve fitness in evolutionary engineering, this strategy can be used for positive selection and rapid identification of genes that are disadvantageous under the given conditions. While this procedure is normally performed with pools of up to a few hundred mutants at a time, hybridization to a high-density array (DNA Chip) of signature tags provides an interesting option for genome-wide selection and identification of relevant genes [50] (see also Sect. 6). Additionally, random insertion-duplication mutagenesis can be used when efficient transformation systems are available [51].

2.5
In Vivo Recombination

Although generally perceived of as clonal, prokaryotes show a wide range of population structures that range from almost strictly clonal (e.g., *Salmonella*) to fully sexual (e.g., certain *Neisseria*) [52]. Akin to directed evolution of proteins, it would be of utmost importance to enhance recombination between different variants with improved phenotypes. To exploit the potential of homologous recombination for evolutionary engineering, DNA exchange within a population may be mediated by the well-known natural mechanisms of horizontal DNA transfer: conjugation, transduction, and transformation. An applied example of this approach is strain improvement of starter cultures in the dairy industry using naturally occurring conjugative plasmids [22]. The use of natural or artificial (e.g., plasmid- or virus-based expression libraries) horizontal DNA transfer and non-homologous recombination, on the other hand, also allows random DNA transfer from other organisms or previously selected variants into a host prior to selection. Thus, appropriate selection will enrich for clones bearing DNA segments that confer a selective advantage and, upon continuation of selection, additional fitness-increasing mutations can occur in this background.

In contrast to the haploid prokaryotes, the use of eukaryotic microorganisms that may exist in haploid, diploid, or even polyploid form, such as *Saccharomyces cerevisiae*, offers the potential for breeding independently improved variants, for instance by creating a diploid cell from two haploids. The offspring from this chimeric diploid cell may than be selected for improved combinations of both haploid variants. This very powerful approach for evolutionary engineering has often been used in industrial strain development of fungal production processes. For example, desired qualities such as robustness, high growth rates, or sporulation have been reintroduced into high yielding, but crippled production strains [34]. It is a pertinent question to ask whether, given the choice, haploid or diploid strains should be used in an evolutionary experiment. It is interesting to note in this context that the frequency at which adaptive mutations are fixed in diploid populations of *S. cerevisiae* was found to be 1.6-fold higher than the frequency in isogenic haploid populations [53]. Although it was argued that diploidy would slow down adaptation under many conditions [54], it appears to be advantageous in asexual populations when the number of favorable mutations per generation is very small – a situation that is not unlikely to occur in evolutionary engineering.

As opposed to the random recombinatorial approaches discussed above, a major benefit to complementing evolutionary engineering with rational design using genetic engineering resides in the potential to jump into new, rationally selected regions of the fitness landscape. Such designs may be based on knowledge of genes or proteins that are anticipated to be relevant for a particular phenotype and this insight would then be used to preselect genes for random expression in selection experiments. Such hypotheses about the relevance of components may be rather vague as hundreds to thousands of genes could be propagated in evolving populations. In practice, rational evolutionary design can be achieved either with multiple heterologous variants of one or more chosen genes or with entire expression libraries of heterologous organisms with desired features. An example of such a rational design is the improvement of recombinant plasmid stability by random cloning of DNA fragments from stable endogenous plasmids [55]. If transfer of large numbers of genes or of entire genomic segments is anticipated, artificial bacterial or yeast chromosomes that allow stable propagation of DNA segments up to several hundred kb in length may replace plasmid-based expression systems.

3 Selection

Natural evolution is thought to be responsible for the extraordinary variety and complexity of the biosphere, and today's life forms are the variants that are presently most fit variants to cope with their particular environments and ecosystems. In the simplest form of directed evolution, a person that differentially removes certain phenotypes from the population establishes relative fitness by screening of individual variants [21, 22]. The obvious advantage of selection by screening is the flexibility that basically any cellular function can be used, provided that a suitable assay is available. Such screening applications profit significantly from recent advances in high-throughput procedures such as robotic (sub-) microliter liquid handling, 384- and 1536-well microtiter plates, digital camera-equipped picking robots, and analytical procedures such as parallel photocells that can rapidly access the various microtiter plate formats. These technical advances are also, in part, responsible for the success of directed evolution strategies in protein engineering. Two general problems pertain to such step-wise evolution approaches: the size of local fitness landscapes for complex cellular phenotypes that require multiple, often unlinked genetic modifications and the strong dependence of phenotypes on environmental conditions. Thus, a critical question is if interesting phenotypes that are identified in multi-well screening procedures translate into the conditions of production processes.

The power of continuous evolution resides in its efficiency and the possibility to select under process-relevant conditions. To avoid unanticipated solutions, the selection procedure should reflect the characteristics of the industrial process, for example aeration, carbon limitation or abundance, fluctuating or constant substrate supply, complexity and concentration of the nutrient sources, pH, osmolarity, mechanical stress, liquid or solid media, cell density, etc. In certain cases, however, pleiotrophic effects of evolutionary adaptation to a particular environment may also increase competitiveness in an alternative environment (see, for example, [56–58]). It needs to be borne in mind that fitness in continuous evolution is a function of competition among the variants that are present under the given conditions, and this property is not under the direct control of the experimenter. Any property that increases the relative number of a variant or the ability of one variant to limit the number of offspring left by other variants under the imposed conditions would improve competitive fitness. Such competitive fitness in a population is not necessarily identical with fitness in the biotechnological sense, which usually refers to improved properties at the single cell level.

3.1 Natural Evolution

The genome of each organism contains not only information for its functioning in the current environment, but the potential to evolve novel functions that will allow it to thrive in alternative environments [19]. To improve understanding of this process and the selective constraints, microorganisms with their short generation times are perfect research subjects, because thousands of generations can thus be studied in simple laboratory environments. At their most basic level, the 'rules' of evolution are remarkably simple: species evolve by means of random variation (via mutation, recombination, or other operators); this is followed by natural selection in which the fittest tend to survive and reproduce, propagating their genetic material to future generations. In addition to horizontal DNA transfer, novel catabolic or metabolic functions are often acquired by mutational activation of cryptic genes, which constitute a versatile genetic repertoire that enhances the adaptive potential of a species [59]. Such cryptic genes are phenotypically silent DNA sequences, which are not normally expressed under any conditions, and are assumed to have played important roles in natural evolution. Another important group of genes in this context are the so-called evolution genes, whose main function in DNA repair appears to be acting for the benefit of evolution itself by generating and modulating spontaneous mutagenesis [30, 31]. Different from mutator genes, however, the rate of mutagenesis that is introduced by these evolution genes is subject to cellular control.

Evolutionary adaptation of species to changing environments occurs in all but the simplest cultivation systems. In fact, our so-called wild-type laboratory strains are the product of an evolutionary domestication process, perhaps most pronounced for *S. cerevisiae*, which has been exploited for baking and alcohol production by virtually every human society. The phenomenon of evolutionary adaptation to laboratory environments has long been recognized and is known as *periodic selection*, referring to the periodic appearance and subsequent exponential take-over of the population by variants with a selection advantage over the currently present cells [60–62]. The kinetics of such population take-overs can be monitored by tracking the replacement of the resident population via markers that have no impact on the fitness of the cells under the cultivation conditions used. This will reveal repeated (periodic) fluctuations in the level of the independent, or neutral, marker. Because these mutations are completely neutral, gain-of-function reversions for such phenotypes, e.g., resistance to a phage or a chemical or utilization of a substrate (other than the one actually used during selection), occur at a constant rate that equals the mutation rate and thus these phenotypes should increase linearly in a population of constant size. In contrast, variants with fitness affecting mutations will substitute the population at a rate that is a function of population structure as well as strength and direction of the selection.

In a culture inoculated from a single clone, a new advantageous mutation is most likely to occur in the much larger population that does not have the neutral mutation, as illustrated schematically in Fig. 3. The adaptive mutant then replaces the currently existing population (including the fraction of neutral mutants) at the log linear rate of selection. The neutral mutation will continue to occur at the same linear frequency in the adaptive mutant, until another advantageous mutation occurs, again in the still predominant population without the neutral marker phenotype. Thus, the abundance of the neutral marker phenotype drops again and the cycle is repeated. Extensive experimental evidence for this phenomenon is given in the excellent review of Dykhuizen [61]. Periodic selection and hitchhiking in bacterial populations are also discussed on theoreti-

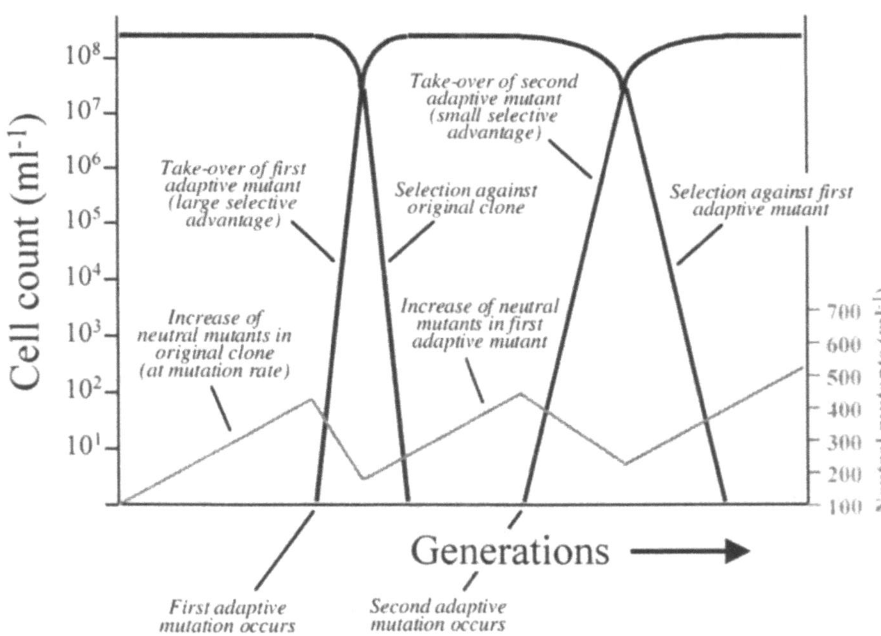

Fig. 3. Schematic representation of the population dynamics during adaptive evolution of an asexual population. The *gray line* at the bottom represents the abundance of neutral mutants (at a linear scale). The other lines indicate periodic selection of two consecutively evolving advantageous mutants (at a logarithmic scale). This was inspired by a similar drawing by Dykhuizen [61]

cal grounds by Berg [60], who developed a stochastic theory to describe the dynamics of large asexual populations.

In addition to monitoring mutant take-overs, such neutral markers are particularly valuable for quantifying differences in fitness between evolved clones. In studies on natural evolution, differences in fitness may depend on subtle variations at one or more loci so that the overall fitness is often difficult to identify. For this purpose, competition experiments are performed using two strains that are distinguished by different neutral markers [61]. By following the relative numbers of two competing strains during a growth experiment, the differential growth rate (s) per unit time (t) can be determined from a plot of $ln(x_i/x_j)$ vs time, where x_i and x_j denote the cell densities of the two strains. Competitive fitness of one strain over another is then quantified by the selection coefficient s_{ij} according to

$$\ln[x_i(t)/x_j(t)] = \ln[x_i(0)/x_j(0)] + s_{ij}t. \quad (2)$$

3.2 Solid Media

Selection on solid media is frequently used because large numbers of mutants can conveniently be screened by visual inspection of growth as such, a zone around the colony as a consequence of a diffusing product, or a color change due to a coupled reaction. Generally, useful results are obtained only when expected differences in fitness are large and the advantageous types are rare. In empirical strain development, plate selection procedures are frequently used for removal of specific feedback inhibition loops in biosynthetic production pathways by selecting for resistance to an antimetabolite of the regulatory substance. The parent strain cannot grow in the presence of this antimetabolite, but any mutant capable of growing must not be feedback inhibited any more [21]. Another example of positive selection for increased tolerance of toxic compounds is the selection for increased antibiotic resistance based on overexpression of inactivating proteins [63].

An advantage of step-wise plate selection is its direct read-out on the progress of evolutionary adaptation, in particular when it is unclear a priori to what extent improvement is possible (see, for example, [64]). However, this mode of selection is likely to be inefficient for complex phenotypes that require multiple mutations. Moreover, the ultimate destination of most strains are some sort of bioreactor, and the importance of mimicking the most relevant production system conditions during selection cannot be overemphasized. From this perspective, plate-based selection assays have an inherent danger of selecting for phenotypes that are not reproducible in liquid media.

3.3 Batch

In liquid media, fitter variants in a particular environment evolve over time and eventually replace the parental population as a consequence of adaptation by selection, which is often studied in batch cultures. An important characteristic of selection in batch culture are dramatic changes in environmental conditions from feast to famine, so that the cells are subjected to alternating periods of growth and stasis upon serial transfer.

A particularly intriguing set of asexual evolution experiments in batch culture was performed by Lenski and coworkers and encompassed the fitness analysis in 12 independent *E. coli* populations founded from a single ancestor [65–67]. Daily serial transfer propagated these populations for 1500 days (about 10,000 generations) in the simple, unstructured environment of glucose-supplemented minimal medium in shaking flasks. After 10,000 generations, the average fitness of the derived clonal variants was increased by about 50% relative to the common ancestor, based on competition experiments in the same batch culture environment. The primary reason for this improvement was attributed to reduced lag phases and higher maximum growth rates. Experiments with alternative carbon substrates also revealed higher fitness on substrates with similar uptake systems, which suggests enhanced transport as an important target of evolution [66]. Although these phenotypic changes were consistent in the 12 independently evolved populations, their genetic diversity – as determined by analysis of restriction fragment length polymorphism with seven insertion sequences as probes – was large [65]. Over time, the evolved genomes became increasingly different from their ancestor and each other, to the extent that almost every individual within a population had a different fingerprint after 10,000 generations. Point mutations were rather rare in the evolved populations, meaning that the accumulated genomic, and possibly phenotypic, changes were mostly a consequence of chromosomal rearrangements. Certain pivotal mutations were apparently shared among all members of a given population, and these constitute prime candidates for phenotypically relevant mutations.

Thus, evolution of adaptive performance is remarkably reproducible, although the phenotypic adaptation may be achieved by greatly different genotypes. While probably only a handful of mutations were relevant for the investigated phenotype, at least some of the other genetic alterations would certainly gain importance under different environmental conditions. Consequently, the history of evolved strains from continuous evolution experiments is very important, as identical selections will inevitably lead to different variants. Another very important observation that pertains to applications of evolution procedures is the hyperbolic rate of change in competitive fitness, as about half of the phenotypic improvement occurred within the first 2000 generations (of 10,000 generations) (Fig. 4). Thus, the rate of fitness gains in microbial populations appears to decelerate significantly over time.

3.4 Microcolonization

A particular problem in selecting for variants with improved secretory capacity in liquid media is the absence of a physical link between the clones in a population and their secreted products. This may lead to interactions between individual clones, such as cross feeding or inactivation of selective agents by few clones within a population. Faced with this problem, a group at Genencor developed an

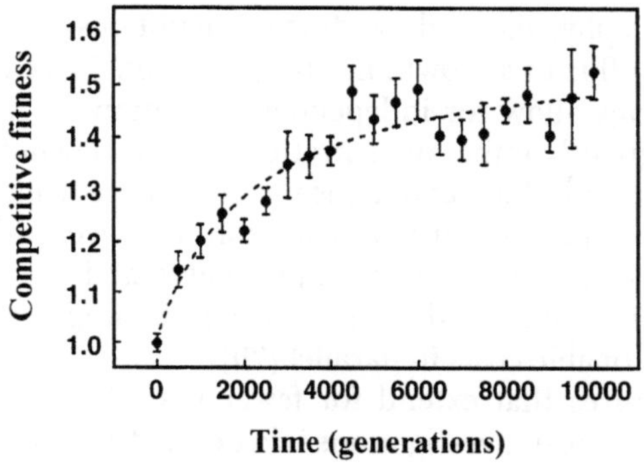

Fig. 4. Change in competitive fitness during 10,000 generations of experimental evolution with *E. coli*. Fitness is expressed relative to the common ancestor. Each point is the grand mean averaged over twelve replicate populations. *Error bars* are the 95% confidence intervals. The *dashed curve* indicates the best fit of a hyperbolic model to the data from Lenski and Travisano [67]. Figure reproduced with permission from Lenski et al. [66]

innovative strategy that enabled the efficient enrichment of better protein secretors from large populations by growing the cells in hollow fibers. The 0.5-μl interior compartments of the fibers act as miniature cultivation vessels [68]. Under these microcolonization conditions, each colony grows in its own microenvironment and cross feeding between neighboring colonies is effectively eliminated. When bovine serum albumin is the sole nitrogen source, clones that secreted either more protease or a better protease variant grew faster than the parent did. After four rounds of selection in such microcolonies, the population was sufficiently enriched with variants exhibiting increased secretion to allow for detailed characterization of individual mutants [68]. Because each hollow-fiber cartridge provides about 3×10^5 such 0.5-μl compartments, this technique is applicable to populations that are too large to be analyzed by screening in microtiter plates. In addition, this procedure can simply be repeated with enriched populations for several rounds such that a bio-panning effect is achieved, which is not possible by selection on solid media. Given its apparent technical simplicity, this approach should also be applicable to other secreted products, provided that a positive selection method can be conceived.

3.5 Chemostat

During growth in batch culture, a population typically passes through the distinct phases of lag, exponential, transition, and stationary growth. Thus, evolutionary events may arise from advantages in any of these phases. In contrast, continuous culture systems provide a constant environment that is also frequently used for studying evolution [61, 69]. Under continuous culture conditions, the removal of cells from the growth chamber by outflow is random and thus becomes a selective function with the growth rate as the main factor determining survival. The most frequently used continuous culture system is the chemostat, which, in physiological steady state, maintains a constant cell density by the continuous influx of a growth limiting nutrient. These well-defined environmental conditions allow for independent variation of growth parameters such as the rate of growth or the concentration of a limiting nutrient. Bioreactors for continuous culture in biotechnological research are usually equipped with sophisticated (and expensive) instrumentation. However, this expense is not necessarily required for evolutionary experiments and the choice of smaller scale chemostats with a simpler design allows performing continuous evolution experiments at reasonable costs in parallel [70].

Continuous cultures that extend for fewer than 20 generations allow for quantitative physiological investigations in a defined steady state. Experiments of longer duration become the study of evolution in action. In continuously operating production processes, the danger of genetic drift resulting from spontaneous mutations poses significant challenges. This is of practical relevance because recombinant organisms are usually engineered to maximize product formation, often at the expense of growth rates or overall fitness. Mutations that increase growth rate will be advantageous and eventually take over the population, thereby likely reducing product formation. However, if used properly, direct control of physiological culture parameters in continuous cultures is a valuable tool that can be employed to modulate selective pressure in favor of a desired phenotype. The influence of these parameters on the competition between different species was reviewed by Harder et al. [36]. When the limiting substrate in a chemostat is the carbon source, the culture is characterized by high efficiency in converting carbon to biomass. When growth is limited by nutrients other than the carbon source, the carbon flux into the cell is generally less tightly controlled, leading to profound effects on cellular energetics [71]. The specific effects of nitrogen, phosphate, potassium, sulfur, and other limitations are reviewed by Dawson [72]. In such cases, various metabolic by-products (e.g., acetate or lactate) or extra- and intracellular polymers are often overproduced, as compared to carbon-limited operation. Consequently, the choice of limiting nutrient will profoundly influence the selection pressure in a chemostat.

During prolonged cultivation in carbon substrate-limited chemostats, two general types of evolutionary events that confer selective advantages to emerging mutants prevail – increased maximum specific growth rates and reduction in the value of the Monod constant K_s for the limiting nutrient [69, 73]. However, any mutation that increases the residence in a chemostat will be favorable, including adherence to bioreactor walls. An important phenomenon concerning the clone-specific metabolism in such evolving cultures is cometabolism, which manifests itself as a physiological and often morphological polymorphism within the population [57, 74]. A particularly well-studied example is *E. coli* cultures in glucose-limited chemostats. A single clone evolved over the period of 773 generations at a dilution rate of 0.2 h^{-1} to form a polymorphic population in which several distinct mutant strains coexisted [74]. In this miniature ecosystem, the largest fraction consisted of efficient glucose scavengers with a metabolite secretion phenotype, and the smaller fraction consisted of mutants that thrived on the secreted, incompletely oxidized metabolites acetate and, to a lesser extent, glycerol [75]. Such an acetate-cross feeding polymorphism is reproducible in long-term populations of *E. coli*, occurring in 6 out of 12 independently studied glucose-limited chemostat populations [76]. In all cases, it was associated with semi-constitutive overexpression of acetyl-CoA synthetase, which allowed for enhanced uptake of low levels of exogenous acetate. Such a polymorphic coevolution potentially complicates selection strategies as the whole population may express a desired phenotype that is not exhibited by any single variant within the population.

Another potential drawback of continuous asexual evolution in continuous culture is the strictly sequential appearance and fixation of adaptive mutations. Consequently, a newly appearing variant may compete only with its immediate one or few predecessors, if historically older variants were previously counterselected. Thus, new variants could in fact exhibit lower fitness compared to more distant predecessors. Such a result was seen with haploid and diploid *S. cerevisiae* cultures that were grown in glucose-limited chemostats for up to 300 generations [77]. As expected, the relative fitness of clones isolated later was always higher than that of the clones isolated immediately preceding the adaptive shift. This was shown by pair-wise competition experiments in which the frequency of the strains was monitored by newly introduced neutral markers. In several cases, however, the relative fitness of clones carrying multiple adaptive mutations were lower than the fitness of clones isolated earlier in the experiment. Thus, combinations of adaptive mutations may result in maladapted clones, as compared to their progenitor, which may have never directly competed with the later occurring variants. During selection in batch culture for 10,000 generations, in contrast, a steady, although hyperbolic improvement in fitness compared to the ancestral strain was observed, as is illustrated in Fig. 4 [66].

The discussions in the previous two paragraphs warrant a note of caution for the use of continuous culture selections in evolutionary engineering of useful phenotypes. Fitness of a particular variant in continuous culture is not only a function of its capability to thrive under the given chemical and physical conditions – usually the phenotype desired by the applied scientist – but is inevitably linked to the presence of and, possibly, interaction with other variants. Thus, fitness in continuous culture is determined by the ability to compete with all other variants that are present at a given time under the applied conditions. This is not necessarily identical with the improvement of a biotechnologically desired phenotype. Because there may not be one optimal phenotype for any set of variants and environmental conditions [60], a population could be cycling through periodic selection indefinitely without actually achieving a long-term improvement in fitness (or a desired phenotype). To ensure that evolutionary adaptation during continuous selection proceeds indeed in the desired direction, it is of utmost importance to monitor evolutionary progress at the single clone level. Additionally, it is probably good advice to inoculate occasionally a new selection culture with the best clone(s) from different stages of the previous selection culture(s), so as to avoid or at least minimize potential evolution of both co-metabolism and unfavorable combinations of adaptive mutations.

3.6 Other Continuous Culture Devices

Variations of conventional chemostats that enable alternative modes of operation for continuous culture have been introduced and exploited. One example is auxostats that modulate the rate of feeding to control a state variable in continuous culture [78]. These devices can be operated under difficult or unstable conditions and thus overcome some of the disadvantages associated with chemostat cultures [78, 79]. Generally, auxostats permit growth near the maximum growth rate without the danger of washout that is inherent to chemostat operation. At high dilution rates, selection rates are remarkable because the effects of small differences between growth rate and washout are magnified. As the culture calls for increased feeding to maintain a constant value of the control variable, there is an accompanying decrease in residence time, which causes slower growing variants to washout. Probably the best known auxostat is the turbidostat, which maintains a constant cell density (turbidity) of an exponentially growing culture using an optical sensor for feedback control of nutrient inflow [80]. A major problem for long-term turbidostat cultivation is microbial adhesion to surfaces, including the optical sensor, as this confounds the turbidity determination. However, the choices of feedback parameters for auxostats are quite broad, including pH, concentrations of dissolved oxygen, nutrients, or metabolic (by-) products in the culture broth, and the concentrations of CO_2, O_2, or volatile compounds in the effluent gas, as well as combinations thereof [78].

Growth in auxostats is usually limited by the availability of a nutrient but may likewise be limited by toxic or inhibitory substances in the growth environment or by some other environmental stress. Generally, variants that are tolerant of toxic agents evolve quickly, and the selective pressure must be increased to further increase the tolerance level and/or to suppress adaptations in which a few members of the population consume or inactivate all the toxin. In the latter case, the selection pressure would effectively be relieved for the rest of the population [81]. To optimize adjustment of the selection pressure, the stress should be increased automatically, preferably via feedback control utilizing a growth parameter that can be measured on-line. Upon periodic mutant take-over, the environmental stress is thus gradually increased in a procedure that is referred to as interactive continuous selection. In principle, any growth parameter could be used for automatic feedback control, provided an appropriate sensor and control design is available.

A particularly ingenious automatic feedback system for interactive continuous selection was devised by Brown and Oliver [82], who used the CO_2 concentration in the effluent gas of a continuous culture to maintain selective pressure for tolerance to increasing concentrations of ethanol in a process that is also referred to as Brown and Oliver interactive continuous selection (BOICS). Specific applications of BOICS are reviewed in Sect. 5.1. Using a model-based approach, guidelines for appropriate BOICS controller design were recently presented that will likely pave the way to a broader application of this very useful selection technique [83]. Comparing the outcome of selection for inhibitor-tolerant mutants in chemostat, turbidostat, and BOICS, it was argued that only the latter selects specifically for variants that are tolerant to extreme concentrations of the inhibitor [84]. Chemostats, in contrast, select for tolerant mutants that can sustain a given growth rate, whereas turbidostats select for tolerant mutants that exhibit increased growth rates under the given nutritional conditions and inhibitor concentrations.

3.7
Fitness Landscapes and Effective Means of Conquering Fitness Peaks

All possible genotypes represent the sequence space, whereas the functional values of the associated phenotypes (or phenotypic characteristics) commonly called fitness, define a fitness landscape. We can conceive of evolution as carrying out adaptive walks towards peaks in more or less mountainous fitness landscapes of sequence spaces, such as among possible DNA or protein sequences. This walk is guided by incremental increases in competitive fitness to drive the distribution of a population towards regions of higher fitness. Although this general view is widely accepted as a fact, quantitative population genetics of adaptive evolution is still a matter of debate [85, 86].

The concept of fitness landscapes as introduced by Wright [16, 17] provides an important contribution to evolutionary theory and is a very useful concept for the discussion of evolutionary processes. Such fitness landscapes are not fixed in structure but deform in response to changes in the abiotic environment and in response to coevolution [15]. In coevolutionary processes, the fitness of one organism depends upon characteristics of another organism with which it interacts, while all simultaneously adopt and change. Although evolutionary engineering is usually initiated with a single strain, coevolution can occur in evolving populations as shown for example in Sect. 3.5. The movement of a population over the fitness landscape depends on the topology of the landscape and on whether the population is sexual or asexual. Local protein-fitness landscapes in directed evolution are usually assumed to be 'Fujiyama-like' (i.e., they increase more or less monotonically towards a fitness optimum) because the protein under investigation has already some characteristics of the desired kind [18]. In contrast, most local fitness landscapes of cellular phenotypes are rugged or, if an organism does not exhibit a desired characteristic (for example utilization of a nutrient), are mostly plain (that is empty of function) with isolated peaks of fitness. For a more comprehensive treatise of this subject, the interested reader is referred to the excellent and provocative book of Kauffman [15].

In general, natural selection tends to drive a population to the nearest peak, which is not necessarily a global optimum. Because there are usually many molecular solutions that enable individuals to surmount environmental challenges, there will be many fitness peaks, the majority of which represent local optima. Depending on whether a population occupies a single niche at high density or is dispersed sparsely over a wide range, it reaches a state of either near-stasis (which most likely represents a local fitness optimum) or gradually improving adaptation, respectively. As microbial laboratory populations are usually of the former type, adapted populations in evolutionary engineering may be stuck with a suboptimal solution to cope with its environment because natural selection opposes passage through a 'valley' of maladapted intermediate states. This theory is, at least partially, supported by Lenski's 10,000-generation experiment, in which resulting populations have seemingly reached distinct fitness peaks of unequal height [66]. In this context, two questions are of immediate applied interest. First, how much time is required for a population to attain a local optimum (or how can this time span be reduced) and, second, how can populations be treated so that they arrive at a global optimum?

The answer to the first question is appropriate tuning of the rate of mutagenesis to minimize the time of selection. Various approaches to that end are covered in Sect. 2. Moreover, it may be advantageous for efficient evolutionary engineering to modify slightly the selection scheme at appropriate intervals. This is because adaptation to the selection conditions usually involves first a modest number of mutations that exert large positive effects that are followed by a greater number of mutations of smaller effect, as was shown both experimentally (e.g., [66]) and on theoretical grounds [85, 87] (Fig. 5). Clearly, it is of utmost importance for any evolutionary engineering experiment to monitor the progress of evolution. Slight modifications in selection schemes may also avoid evolution of overly specialized variants that exhibit the desired phenotype only under the exact conditions of the selection. The answer to the second question is recombination, so that a population does not necessarily need to reinvent novel properties, as they could simply be transferred from different organisms or previously selected variants. Selection is then used to choose the most appropriate from different molecular incarnations of this property and to incorporate it optimally into the host strain.

While the above discussion concerned crossing of valleys between different but related fitness peaks, another problem is the distance between the starting point in sequence space and the nearest fitness peak. This poses the practical

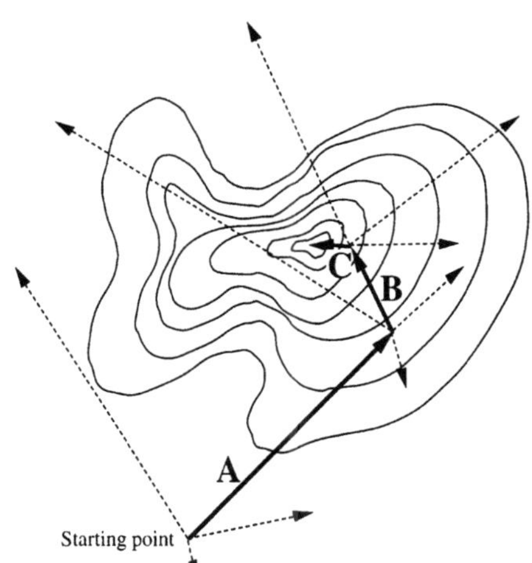

Fig. 5. An evolutionary walk to the optimum in a three-dimensional fitness landscape. The *arrows* represent random mutations having different magnitudes (length) and directions (effect on fitness). *Solid and dashed arrows* illustrate beneficial (A to C) and ineffective/detrimental mutations, respectively

difficulty of achieving multiple mutations to yield any improvement in the desired phenotype, in particular for evolutionary engineering of novel phenotypes. Consequently, there may not be a gradually ascending slope to the nearest fitness peak for guiding the evolutionary walk. A practical example is the requirement of three novel enzyme activities to convert a non-metabolizable nutrient source into a common biosynthetic intermediate. In this case, there is no increase in fitness if only one or two of these enzymes become available. Therefore, even in the most advantageous scenario where the required enzymes are already present in the form of cryptic genes, chances for simultaneous appearance of three independent deregulatory mutations in one variant are very low (6.4×10^{19} for the case of three independent point mutations in a genome with 4000 kb). In such cases, evolutionary approaches are likely to fail unless extremely large populations or rationally selected pathway intermediates are used (see also Sect. 4). Nature approaches this problem by recombination and horizontal DNA transfer (see Sect. 2.5), which allows 'jumping' closer to a fitness peak. For certain phenotypes, such DNA sequences may have to be provided by the experimenter.

Naturally it would be desirable to predict the success of selection schemes. Although, in many cases, this may not be possible with any confidence, some general guidelines may be given. The chances of selecting a phenotype of interest in a particular organism are good when (i) a phenotype can be detected in at least rudimentary form, (ii) a fairly close relative of the organism in question exhibits the phenotype, (iii) a related phenotype such as activity toward an analog of a novel substrate can be detected, or (iv) important aspects of the phenotype are susceptible to recombinant approaches because they are encoded on transferable genetic elements such as a few genes or operons.

3.8
Screening of Desired Variants from Evolved Populations

According to the quasispecies concept, the result of evolution is not a single variant, but rather a distribution of related variants that occupy a distinct region in sequence space [12]. Consequently, populations evolved from continuous selections are often heterogeneous, and representative, often large, numbers of individual clones from such populations must be examined to identify the most suitable individuals. The most important prerequisite for screening is efficient spatial separation and access to an assay system that allows characterization of the desired phenotype. To this end, several methodologies with different levels of automation and throughput are presently available [20].

The highest throughput can be achieved by the combination of flow cytometry and cell sorting. This is a rapid method for the analysis of *single* cells as they flow in a liquid medium through the focus of a laser beam surrounded by an array of detectors. By simultaneous use of different fluorescent stains, flow cytometry can yield multiparametric data sets which are, however, often difficult to interpret [88]. These are then used to discriminate between different types of cells, a procedure that is suitable for rapid enrichment of certain types of cells from large populations. An important and potentially very useful contribution to fluorescence-based screening comes from green fluorescent protein and its recombinant derivatives, which can also be exploited as expression markers at the single cell level.

Most analytical methodologies, however, cannot function at the single cell level. This means that variants have to be characterized as cultures, which requires laborious segregation, isolation, and cultivation of individual clones. In the simplest case, a desired phenotype is defined by growth under certain conditions, so it can be directly assessed by visually inspecting the ability to grow on plate or in liquid media. However, desired growth phenotypes frequently cannot be determined by a simple yes or no experiment, but are based on improved tolerance of certain unfavorable process conditions, in which case survival becomes a statistical process. In such cases, the survival rate is usually estimated by comparing colony-forming units on solid media. Alternatively, survival can also be assessed by measuring the most probable number of viable cells, based on the potential of various dilutions of the culture to serve as an inoculum for liquid media [89]. In practice, three to five serial dilutions are performed in parallel and used as inocula in a procedure that readily lends itself to automation in microtiter plates [90]. A great deal of ingenuity has also gone into the design of protocols that couple a desired function with activation of a marker gene, which than effects a color change if used with appropriate chromogenic substrates [8].

Additionally, a variety of analytical equipment and techniques that allow the examination of small- (and micro-) scale microbial cultures and their products have become available. Examples include near infrared and Fourier transform infrared spectroscopy, which offer the ability for in situ detection of specific compounds in fermentation broth [22]. However, sensitivity and the required sample volumes pose serious obstacles that still have to be overcome. Another alternative is offered by sensitive pyrolysis mass spectroscopy, which was demonstrated to be suitable for quantitative analysis of antibiotics in 5-µl aliquots of fermentation broth when combined with multivariate calibration and artificial neural networks [91]. The authors concluded that a throughput of about 12,000 isolates per month could be expected. Furthermore, standard chromatographic methods such as gas chromatography or high-performance liquid chromatography, possibly in combination with mass spectroscopy (MS) for detection, can provide simultaneous quantitative detection of many metabolic products.

Given the availability of analytical procedures, throughput is now largely limited by the ability to cultivate cells in suitably miniaturized vessels that provide process-relevant environmental conditions. Although many microbes are, in principle, amenable to growth in microtiter plates, investigation of their phenotypes in the standard 200-µl working volume plates is often limited to qualitative information because aeration and/or mixing tend to be limiting [92]. An interesting alternative is a recently developed miniaturized microbial growth system that consists of special 96-well plates equipped with deep (2-ml) wells and a spongy silicone/cotton wool sandwich cover that adequately prevents both cross contamination and excessive evaporation during vigorous aeration [93]. It was shown that aeration in these deep-well microtiter plates was comparable to that in baffled shake flasks and allowed the attaining of cell densities of up to 9 g dry weight per liter. Such cultivation systems in combination with appropriate analytical tools will enable quantitative physiological characterization of larger numbers of clones.

Data from such characterization studies may then also be used for metabolic flux analysis, a method of estimating the rates of intracellular reactions. This modern offspring of quantitative physiology combines data on uptake and secretion rates, biosynthetic requirements, quasi-steady state mass balances on intracellular metabolites, and assumptions about metabolic stoichiometry to compute the intracellular flux distribution [94]. In addition, ^{13}C-labeling experiments are now increasingly used to avoid or validate critical assumptions [95]. Currently, labor and expense prevent the direct application of such methodologies in screening processes, but less complex approaches may offer the possibility of examining intracellular flux responses at reduced resolution in a smaller-scale screen [96]. For example, using a recently introduced nuclear magnetic resonance methodology based on isotopic imprinting of amino acids by their precursors, the active central carbon pathways and the ratios of their fluxes can be directly determined from two-dimensional nuclear magnetic resonance analysis of ^{13}C-labeled biomass [97]. This metabolic flux ratio analysis was recently demonstrated to provide valuable insights into intracellular carbon metabolism of different *E. coli* strains under various environmental conditions, including shake flask cultures [98]. Further increases in throughput can be expected from the use of MS-based procedures for labeling pattern analysis [96, 99, 100]. The interest in metabolic flux analysis resides in its analytical power at the metabolic level and its potential to provide insights for strain improvement, genetic manipulation, and process optimization. Thus, the growing field of metabolic flux analysis together with functional genomics [101] and computational models of cellular metabolism [102, 103] will likely become important tools in directing screening work, possibly by identifying easy to determine physiological variables that are indicative of a desired phenotype.

4
Evolutionary Engineering of Simple Cellular Subsystems

Evolutionary selection principles have been used to approach biotechnological problems of various complexities (Table 2). In the simplest case, conceptually, a desired phenotype is based on a 'single property' and is thus susceptible to straightforward gain-of-function selection. In such cases, the behavior of a relatively simple cellular subsystem (e.g., transport of a nutrient) can be directly linked to fitness in the selection scheme. In the definition employed here, simple cellular subsystems have only a small, defined number of involved components and, more importantly, their interaction with other aspects of cellular metabolism are not limiting for the property under investigation. For practical reasons, complex cell systems in industrial strain development such as entire biosynthetic pathways are often separated into simpler subsystems. This can be achieved, for example, by selecting for properties that render individual enzymes of such pathways insensitive to toxic structural analogs of pathway intermediates [20, 22]. In the absence of complete knowledge of what components are

Table 2. Recent examples of evolutionary engineering

Evolved phenotype	Selection system	Reference
Novel catabolic activities		
Utilization of carbon substrates (coryneform bacteria)	Plates (with limiting amount of yeast extract)	[114]
Utilization of pentoses (*E. coli*)	Plates (non-growing cells)	[111]
Novel esterase activities (*P. putida*)	Plates (non-growing cells)	[38]
Galactitol dehydrogenase (*Rhodobacter*)	Chemostat (glucose-limited, excess galactitol)	[115]
PTS-independent glucose uptake	Chemostat	[106]
Improved enzyme properties		
Secretability	Microcolonies	[68]
Thermostability	Thermophilic hosts	[8]
Functionality (*E. coli* mutator strain)	Batch (increasing antibiotic concentrations)	[42]
Improved plasmid functions		
Stability (Gram positives, yeast)	Chemostats (antibiotic and auxotrophic marker selection)	[55, 81, 125, 126]
Stable host-plasmid combinations (*E. coli*)	Chemostat	[128]
Improved stress resistance		
Acetate tolerance (yeast)	Turbidostats	[118]
Organic solvent tolerance (mutator strains)		[119]
Ethanol tolerance (yeast)	BOICS	[82]
Antibiotic resistance (*Streptomyces*)	BOICS	[25]
Multiple stress resistance (yeast)	Chemostats and batches (with stress challenges)	[90]
Membrane protein overexpression (*E. coli*)	Plate	[124]
Periplasmic protein production (*E. coli*)	Chemostat	[57, 137]
Improved production properties		
Endo-enzyme overexpression	Chemostats	[109, 110]
Antibiotic production (*Streptomyces*)	BOICS	[25]
Nucleoside secretion (*E. coli*)	Chemostat (phosphate-limited, added biosynthetic inhibitors)	[121]
Protein secretion (*Streptomyces*)	Chemostats (different selection schemes)	[125]
Biomass yield (yeast, *E. coli*)	Chemostat (carbon-limited)	[57, 133, 134]
Adhesive cells (*Streptococcus*)	Chemostat	[108]
Altered mycelial morphology (fungi, actinomycetes)	Chemostats	[125, 129, 131, 132]

involved, however, a priori classification of phenotypic properties according to their complexity is difficult.

A particularly well-studied example of a simple subsystem in evolutionary research is utilization of lactose, which consists of three essential components: (i) porin-mediated diffusion through the cell wall, (ii) active uptake via a permease, and (iii) intracellular hydrolysis into glucose and galactose by β-galactosidase. Assuming that central metabolism will utilize these cleavage products, the lactose flux should be directly proportional to the growth rate in lactose-limited media, and this is indeed the case [104]. In lactose-limited chemostats, periodic selection of *E. coli* predictably generates lactose-constitutive variants [69]. Further beneficial mutations reduce the K_s value of the permease; this is in agreement with the calculated control coefficients for the three components under these conditions [105].

Excluding classical mutagenesis and selection on solid media, there are several reports on evolutionary engineering of simple cellular subsystems with an applied background. For example, experiments were performed with an *E. coli* strain that produced an aromatic compound and carried a deletion of the phosphotransferase system (PTS) for glucose uptake. Spontaneous glucose revertants were selected that apparently utilized a non-PTS system for glucose uptake [106]. One variant was identified that exhibited improved production of aromatic compounds, presumably because the use of a non-PTS uptake system for glucose uptake saves at least some intracellular phosphoenolpyruvate (which is otherwise converted to pyruvate during PTS transport of glucose), increasing its availability for biosynthesis of aromatics. Interestingly, using the same approach in a similar host but following the rational strategy of cloning a heterologous, non-PTS system for glucose uptake did not improve production of aromatics [107]. This example illustrates the advantage of evolutionary engineering for optimally accommodating a metabolic component into the complex system of cellular metabolism. Selection procedures have also been used to improve more specialized desirable properties such as improved downstream processing characteristics or resistance to phage infection. Although usually undesired, adhesive phenotypes can be selected for the use in certain types of bioreactors that require attachment of cells [108].

The isolation of mutants overproducing endo-enzymes that directly influence growth fitness has often been achieved using chemostat selection (e.g., [109, 110]) or other means [111]. A successful example of the conceptually more difficult improvement of exo-enzyme production involves the enrichment of more efficiently secreted protease variants by using bovine serum albumin as the sole nitrogen source in a selection procedure based on microcolonies (compare with Sect. 3.4) [68]. Specifically, (rare) protease variants with up to fivefold increased secretion levels were isolated after mutagenesis and four rounds of selection by growth in hollow fibers. While this strategy was successfully applied to select for better protein secretion, it could also potentially be used to select for host strains that exhibit an improved secretion phenotype. In several cases, evolutionary engineering of thermostable enzyme variants was successfully achieved by expression in thermophilic organisms and selection of transformants for recombinant activity-dependent survival at elevated temperatures (for a review see [8]). This powerful concept may also be extended to microbes capable of growing under other adverse environmental conditions, including extremes of pH and salinity.

Acquisition of novel catabolic activities has been deliberately studied since the early 1960s and is of particular applied relevance for bioremediation of waste or by-products from manufacturing processes and improving the ability to use cheaper raw materials in the production of commodity chemicals. Most of these studies are either conducted with well-characterized laboratory strains [111, 112] or based on the analysis of naturally evolving species in the environment that can degrade pollutants of human origin [112, 113]. When multi-step catabolic pathways are required to degrade a pollutant, the most important mechanism for expanding the metabolic capabilities appears to be incorporation of existing genetic material via horizontal DNA transfer. However, less complex alterations for acquisition of new activities can also be achieved by test tube evolution with a single strain. Such evolutionary gain-of-function selections revealed the general principle that new metabolic functions are often established by 'borrowing' enzyme or transport activities from preexisting pathways [111, 114]. Two types of mutations are found to account for most newly evolved pathways: (i) the initial events are almost always activation of cryptic genes or regulatory mutations of genes normally used in other metabolic pathways, and (ii) subsequent mutations in structural genes that alter properties such as substrate specificity. To select for mutants that can use or degrade new compounds, microorganisms are placed in media containing these non-metabolizable nutrient sources. Typically, cells are provided with a limiting concentration of a normal nutrient to support some growth in liquid or on solid media, because the desired mutants are often not obtained by direct selection [114]. Moreover, it may not be possible to select directly for a desired phenotype in one step when multiple mutations are required. In such cases, it is worthwhile to attempt selection on structural analogs of the novel substrate or intermediates of the anticipated catabolic pathway.

Successful evolution of novel catabolic functions has been demonstrated in a number of bacteria [112]. Using a plasmid-based mutator gene, novel esterase activities were selected in *Pseudomonas putida* [38]. Another application is selection of the 'new' catalytic activity of a galactitol dehydrogenase by cultivating *Rhodobacter sphearoides* in a chemostat with a limiting concentration of a normal substrate and an excess of the non-metabolizable galactitol [115]. After about 50 days, a spontaneous several-fold increase in cell density indicated an adaptive mutation that enabled utilization of galactitol. Biochemical characterization of the resulting galactitol dehydrogenase showed it to be a previously unrecognized enzyme in the wild-type. Evolution of this 'new' enzyme was presumably based upon activation of a cryptic gene (compare with Sect. 3.1). After up to 60 days in stationary phase, mutants capable of utilizing several novel carbon substrates were obtained from industrially important coryneform bacteria that were plated on mineral media with a very low concentration of yeast extract and a high concentration of the carbon source of interest [114]. Alternatively, selection may also be achieved without an initial growth promoting substrate, as evidenced by the isolation of ribose-positive *E. coli* mutants after 12–20 days of incubation in a minimal medium containing ribose as the sole carbon source [111]. The latter two cases of evolutionary adaptation presumably take advantage of the increased rate of mutagenesis and population dynamics during prolonged nutritional stress in stationary phase [29, 116, 117].

Clearly, evolutionary engineering of simple cellular subsystems is complementary but also competing with directed in vitro evolution, provided sequence information on the involved components is available.

5 Evolutionary Engineering of Complex Cellular Subsystems

5.1 Resistance to Environmental Stress

Although modern process equipment enables tight control of many environmental factors, industrial microorganisms often have to cope with adverse conditions that are inherent to an industrial process, for instance high concentrations of toxic or inhibitory products. In many cases, evolutionary procedures have been used to improve performance by adapting strains to such process conditions. For example, moderately acetate-tolerant baker's yeast variants were selected in turbidostats to improve the dough raising power in acetate containing sourbread [118]. Similarly, improved organic solvent resistant bacteria were selected by using mutator strains [119]. Also, to maintain the extraordinary resistance to high concentrations of acetate in industrial acetic acid bacteria that are used for the production of vinegar, these cultures are continuously propagated in acetate fermentations [120]. To avoid problems of over- or under-addition of toxic agents in the selection of mutants tolerant of extreme environmental stresses, the selection pressure is best adjusted automatically in response to periodic mutant take-overs via feedback control of the culture conditions in a process known as interactive chemostat selection (see also Sect. 3.6). In a particular interactive chemostat procedure using CO_2 output as a measure of the culture condition (BOICS), ethanol-tolerant yeast mutants were successfully isolated [82]. BOICS was also used to obtain *Streptomyces griseus* mutants that exhibited greatly increased resistance to the antibiotic streptomycin [25]. Associated with increased resistance, the best mutant produced 10 to 20 times more streptomycin when grown in the medium used for BOICS. The strategy apparently implemented by BOICS uses the mean specific growth rate of the culture as a measure of its health and CO_2 output is used as a measurable surrogate for growth rate to control the environmental conditions [84].

Resistance to inhibitors added to liquid media may also be used to select for variants that secrete desired metabolites, as exemplified by chemostat selection of *E. coli* mutants secreting thymidine, cytosine, uracil, guanine, and thymine [121]. Since it was not possible to favor directly secretion of the desired compound, thymidine, a chemostat population was challenged with increasing concentrations of two inhibitors of the pyrimidine biosynthesis pathway. Phosphate limitation successfully prevented growth disadvantages due to squandering of critical resources under carbon limitation. Thymidine-secreting mutants were then detected on the basis of cross feeding of an auxotrophic *thyA* mutant in a plate assay. Interestingly, the isolated mutants also secreted other nucleosides and nucleobases, so that the underlying principle of this design may be generally applicable to select metabolite-secreting mutants.

Another biotechnologically desirable characteristic of process organisms is robustness or resistance to the multiple stresses that frequently occur in large-scale processes or in food applications. However, increased tolerance of multiple stresses is likely to be a complex phenotype that would be difficult to engineer rationally. A recent study compares selection procedures to select for improved multiple stress resistant phenotypes from chemically mutagenized *S. cerevisiae* [90]. Specifically, glucose-limited chemostats with either permanent or transient stress challenges as well as repeated cycles of mutation and selection against various stresses in batch culture were investigated. Evolution of stress resistance was followed by monitoring the relative tolerance to four stresses: ethanol, rapid freezing, oxidation (H_2O_2), and high temperature. The analyzed samples were either from population aliquots that originated at various stages of the selection processes or, in selected cases, from 24 representative clones that were picked from plates. The most appropriate strategy for obtaining multiple stress resistant variants appeared to be selection in chemostats with transient stress challenges, after which the population was allowed to recover for several generations. Several clones from this heterogeneous population exhibited five- to tenfold improved resistance to three out of the four stresses. Two to three cycles of transient exposure to stresses prior to growth in batch culture, on the other hand, selected for variants with higher resistance (up to 150-fold) but to only two out of four stresses.

5.2 Resistance to Metabolic Stress

Generally, overproduction of antibiotics, vitamins, or fine chemicals constitutes a metabolic and energetic burden for the cell, and hence is frequently counterselected in production processes if not maintained by strong selective pressure [112]. However, even in the presence of marker gene-based selection pressure, a complex phenotype such as vitamin production may be counter-selected during moderately extended cultivation [122].

Another biotechnologically relevant stress stems from toxic effects of recombinant protein overexpression that impair growth of the host cell. While *E. coli* is a powerful vehicle for the overproduction of many heterologous proteins, certain proteins cannot be expressed at all or only at very low levels. Foremost among those are membrane proteins that are difficult to overexpress in both microbial and eukaryotic hosts [64]. This problem may be partly related to the observation that laboratory strains are generally not well suited for protein overproduction, as they have been selected for maximum growth [123]. In a very interesting study, Miroux and Walker [124] provided a solution by selecting *E. coli* mutants that proved to be superior to the parental strain for overexpression of problematic globular and membrane proteins. The plate-based selection procedure was initiated with a strain carrying an inducible expression plasmid for the least toxic of seven tested membrane proteins. After growth and a short induction phase in liquid medium, transformants were diluted on plates containing both ampicillin and IPTG for plasmid maintenance and induction, respectively. Two (minor) sub-populations with different colony sizes survived, one of which had apparently lost the capacity to express the recombinant protein, while the other expressed appreciable amounts of the membrane protein. An isolate of the latter population, morphologically characterized by a small colony size, was found to be a suitable host for overexpression of many previously problematic proteins. Because the toxicity of overexpression for certain proteins persisted in the isolated mutant, a second round of selection was conducted on this mutant after transformation with an expression plasmid for one of the remaining problematic proteins. One of the mutants obtained from this second selection proved to be a better producer for some but not all of the problematic proteins, even compared to the previously isolated mutant. Both mutant phenotypes were stable propagated and are apparently caused by genomic mutations that were hypothesized to reduce the level or activity of T7 RNA polymerase, and so prevent uncoupling of transcription and translation [64, 124].

5.3 Plasmid Stability

Structural and segregational stability of plasmids is a prerequisite for development of efficient processes and, moreover, important for validation of pharmaceutical manufacturing processes. Segregational instability occurs when a plasmid-bearing host fails to pass the plasmid on to a daughter cell(s), and a variety of (often unknown) factors contribute to segregational stability. To improve plasmid retention in Gram-positive bacteria, selective chemostats have successfully been employed to alter both host [81] and plasmid [55] factors. In both cases, cultures hosting segregationally unstable plasmids were grown for up to 100 generations in carbon-limited chemostats at a high dilution rate (of about 0.5 h^{-1}) under selective pressure from supplemented antibiotics. Variants of a normally unstable recombinant *Bacillus* strain exhibiting about 30-fold improved plasmid retention were enriched by this procedure [81]. In this case, the stability characteristics resided in the host rather than on the plasmid. The improved strains had growth rates comparable to that of the original, plasmid-free host and were consequently better competitors. Using a recombinatorial approach, Seegers et al. [55] selected stable plasmids in lactobacilli from a large background population of recombinant plasmids with different stabilities. After shotgun cloning of DNA fragments from a stable lactococcal plasmid into an unstable expression vector, three classes of mutations were selected and subsequently identified. The first class mutations in the selection plasmid itself increased copy number, thereby rendering the plasmid more stable. The other two classes were based on the insertion of two different stability-promoting sequences in the selection plasmid.

In another evolutionary approach, expression and secretion of a recombinant protein in the Gram-positive bacterium *S. lividans* was increased 60- to 100-fold, most likely by improving plasmid stability in combination with other host properties [125]. Improved strains were selected from four consecutive chemostat processes run at a dilution rate of 0.12 h^{-1} under different selection regimes. In the first step, after about 100 generations under ammonium limitation and glucose excess, variants with about fivefold improved recombinant protein secretion were isolated. In the second step, cultivation under maltose limitation for another 100 generations was supposed to lead to increased segregational plasmid stability and clones with 30-fold higher protein secretion relative to the original strain were isolated. Finally, two more rounds of selection with increasingly selective antibiotic concentrations for about 33 generations each were performed, leading to clones that exhibited about 60- to 100-fold increased recombinant protein secretion, as compared to the original strain.

A critical factor for successful selection of segregationally stable host-vector combinations is the selection pressure applied. While the above positive selections for antibiotic resistant cells were successful, a similar experiment that used a negative selection for plasmid-bearing clones of *S. cerevisiae* with an auxotrophic marker did not enrich for more stable clones over a period of 420 generations [126]. Although a large variety of clones with altered recombinant plasmid stability evolved over time, it appeared to be mainly a result of non-specific periodic selection. Moreover, the best clones exhibited only about a 30% improvement in stability. This apparent absence of selection pressure for stable clones may have been caused by cross feeding of the plasmid-free population with the auxotrophic nutrient that was synthesized by the plasmid-bearing population. This is a common phenomenon in recombinant yeast cultures [127]. Similarly, during selection for plasmid retention with chloramphenicol, the selection procedure also promoted a higher rate of chloramphenicol degradation, which, in turn, resulted in a progressive increase of the chloramphenicol-sensitive, plasmid-free population [81]. However, in this case the selection pressure was monitored and could be gradually increased simply by raising the antibiotic concentration.

Although generally considered to impose a burden and thus to reduce fitness, plasmid retention may become beneficial for coevolved hosts by unexpected means. After propagation of a plasmid-carrying *E. coli* strain for 500 generations, a host phenotype evolved that, relative to its progenitor, exhibited a competitive advantage from plasmid maintenance in the absence of selection pressure [128]. Although the mutation within the host genome remained unknown, it was shown that the plasmid-encoded tetracycline resistance, but not the chloramphenicol resistance, was required to express this beneficial effect. These results indicate that the co-evolved host phenotype acquired some new (unknown) benefit from the expression of a plasmid-encoded function. This also suggests a general strategy for stabilizing plasmids in biotechnological applications by evolutionary association of plasmids with their hosts. Thus, antibiotic selection could be avoided in industrial processes without the danger of phenotypic instabilities due to plasmid loss.

5.4
Mycelial Morphology

Mycelial morphology is an important process variable in fermentations with filamentous fungi. This is particularly true for the commercial production of the Quorn myco-protein, a meat substitute with a texture that is based on the morphology of the mycelium. Continuous-flow production of this material by the fungus *Fusarium graminearum* is prematurely terminated if highly branched mutants appear in the process. From a series of glucose-limited chemostats, it was possible to isolate mutants in which the appearance of such highly branched mutants was significantly delayed, compared to the parental strain [129]. A more detailed analysis of periodic selection within the evolving population during continuous production of Quorn revealed that pH oscillations or a consistently low pH are complementary conditions that delay the appearance of the undesired, highly branched mutants, without affecting the normal morphology of the mycelium [130].

For other applications, mycelium formation is undesired and may be reduced by appropriate selection procedures. This was achieved, for example, in the bacterium *S. lividans* by extended growth in chemostat cultures under ammonium limitation and glucose excess [125]. After about 70 generations, selected variants showed an altered growth behavior that was characterized by repression of aerial mycelium and spore formation on solid media. Similar results were obtained with different fungi [131, 132].

5.5 General Physiological Properties

While novel reactions and pathways can often be efficiently installed in microorganisms by metabolic engineering [1], general physiological properties such as specific growth rate, overall metabolic activity, energetic efficiency, competitive fitness, and robustness in industrial environments remain mostly the property of the chosen host organism. It would, therefore, be advantageous if host organisms could be tailored for the specific requirements of different industrial processes. One such industrial example is (*R*)-lactate production with *Lactobacillus* by BASF [112]. In this case, an improved, fast growing mutant was isolated from semi-continuous fermentation in production scale because lactate production is linked to growth.

High yields of biomass represent a general host property that is desired in many applications, and has been achieved by evolutionary strategies. Comparing an *S. cerevisiae* mutant isolated after 450 generations in a strictly glucose-limited chemostat at a dilution rate of 0.2 h^{-1} with its ancestor, Brown et al. [133] found the evolved strain to exhibit significantly greater transport capacity and also enhanced metabolic efficiency in processing of glucose under these conditions. The evolved strain had acquired the remarkable capability to grow at a biomass yield of 0.6 (g/g), compared to 0.3 (g/g) for the parent. This improved growth phenotype under strict glucose limitation apparently did not compromise the performance under non-limiting conditions in batch cultures. In fact, the overall yield of cells on glucose was increased in batch culture as well. The two- to eightfold faster glucose uptake of the evolved strain, compared to the parent, was correlated with elevated expression of the two high-affinity hexose transporters, *HXT6* and *HXT7*, which, in turn, was caused by multiple tandem duplications of both genes [133]. Although the genetic basis for the enhanced glucose transport has been unraveled, these genetic alterations are probably not responsible for the biotechnologically relevant phenotype of more efficient biomass production. Inoculated from the same parent, three *S. cerevisiae* mutants were isolated from independent glucose-limited chemostat cultures after 250 generations and all of them produced about threefold greater biomass concentrations in steady state [134]. Reduced ethanol fermentation and increased oxidative metabolism apparently achieved this improvement in metabolic efficiency. Analysis of total cellular mRNA levels revealed significant changes in the transcription levels of several hundred genes compared to the parent, but a remarkable similarity in the expression patterns of the three independently evolved strains [134]. Consistent with the observed physiology, many genes with altered transcription levels in all three strains were involved in glycolysis, tricarboxylic acid cycle, and the respiratory chain. These results indicate that increased fitness was acquired by altering regulation of central carbon metabolism, because only about five to six mutations were expected to contribute to the changes. Possibly as a consequence of the evolutionary principle that different populations may evolve under identical conditions, a different outcome was seen in an earlier but apparently identical selection experiment for 260 generations [135]. In this case, the biomass yields of isolated yeast clones fluctuated with the progress of evolution and clones from later generations exhibited significantly reduced yields under the selection conditions, whereas the yields in batch culture were not affected.

In an effort to select for variants that would perform well under the typical industrial fed-batch condition of slow growth, an *E. coli* mutant was isolated after 217 generations from a glycerol-limited chemostat that was operated at the very low dilution rate of 0.05 h^{-1} [57]. Like the yeast strain described above, this mutant was found to exhibit an increased biomass yield. Additionally, other general physiological properties such as the specific growth rate and resistance to a variety of stresses were found to be improved. Unexpectedly, the mutant also exhibited high metabolic activity in the absence of growth, which indicated impaired stationary phase regulation [136]. Some of these improvements were also evident with carbon sources other than the one used during selection, indicating that not only substrate-specific features but also general physiological properties were altered. In subsequent studies, these improved phenotypic properties were shown to be exploitable for biotechnological applications, including periplasmic secretion of recombinant protein [137] and production of low molecular weight biochemicals [136]. Moreover, the isolated mutant was shown to be significantly less impacted by periplasmic expression of the recombinant protein, as evidenced by the significantly higher segregational stability of the expression plasmid during growth in non-selective media (Fig. 6). Consistent with the total cellular mRNA data obtained from the metabolically more efficient yeast strains, several proteins involved in central carbon metabolism were found at significantly higher levels on two-dimensional protein gels from the isolated *E. coli* mutant [138].

The above examples clearly illustrate that it is feasible to select for generally improved microbial phenotypes for industrial applications. Dictated by economic pressure, it is, however, often impractical to switch host strains in advanced stages of process development. Thus, it would be highly desirable to develop production hosts for the specific requirements of bioprocesses by metabolically engineering them to have desirable physiological properties, which necessitates elucidation of the genetic basis of these often complex phenotypes. In the case of the *E. coli* mutant, this has partly been achieved by identifying two genes, *rspAB*, which, when overexpressed in wild-type *E. coli*, partly mimic the

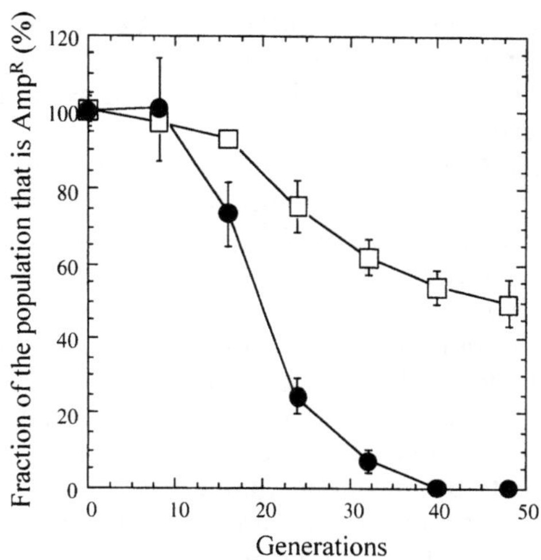

Fig. 6. Fraction of ampicillin-resistant clones of *E. coli* MG1655 (*circles*) and a chemostat-selected descendant (*squares*) from serial batch cultivations in ampicillin-free minimal medium. Both strains harbor the expression vector pCSS4-p for periplasmic production of the recombinant α-amylase of *B. stearothermophilus*. Reproduced with permission from Weikert et al. [137]

mutant phenotype [139]. Specifically, co-overexpression of RspAB was found to improve the formation of recombinant β-galactosidase in batch and fed-batch culture of *E. coli*. Although the exact functions of the corresponding gene products are not fully elucidated, they are reported to be involved in the degradation of the metabolic by-product (or signaling molecule) homoserine lactone [140].

6 Outlook

The use of evolutionary principles will undoubtedly play a major role in twenty-first century biotechnology [141]. The capabilities of directed in vitro evolution will eventually extent beyond improving existing properties of proteins or short pathways to the engineering of de novo functions, new pathways, and perhaps even entire genomes [12, 13]. However, the problem of phenotypic complexity will shift the limitations even more to the available screening or selection procedures [11]. For two primary reasons, evolutionary engineering of whole cells offers an interesting alternative. First, through the use of continuous evolution using large populations, evolutionary engineering can navigate rugged fitness landscapes much more efficiently than can step-wise screening or selection procedures. Second, cellular phenotypes depend strongly on the environment and appropriate process conditions may be simpler to establish in bioreactor systems than in Petri dish- or microtiter plate-based screening or selection systems. Moreover, for complex microbial phenotypes with many, often unknown molecular components, there is currently no alternative to evolutionary engineering. Although such applications were not covered here, evolutionary studies with microbes are also likely to provide important input to medicine, for example by suppressing the emergence of novel pathogens through environmental controls, reducing virulence reacquisition of live vaccines, or avoiding the evolution of drug resistant variants [19].

The greatest limitation for evolutionary engineering of industrially useful cellular phenotypes resides in the contradictory selection demands for such phenotypes. In highly engineered production strains, for example, it may not be possible to devise a selection scheme for two useful but potentially incompatible phenotypes such as overproduction of a metabolite *and* high efficiency of growth. In such cases, both direct evolution and evolutionary engineering approaches are envisioned to become components in effective metabolic engineering, as illustrated in Fig. 7. Upon successful evolutionary engineering towards one desired phenotype, this strain is used either as the host for further rational improvements by metabolic engineering or the desired property is transferred to a production host. The latter is essentially *inverse metabolic engineering*, a concept introduced by Bailey et al. [4]. Here a desired phenotype is first identified and/or constructed and, upon determination of the genetic or environmental basis, it is endowed on another strain or organism.

Until very recently, searching for the genetic or molecular basis of complex phenotypes would have been a hopeless venture because multiple, random genetic changes at the genome level could not be identified. To a large extent, this

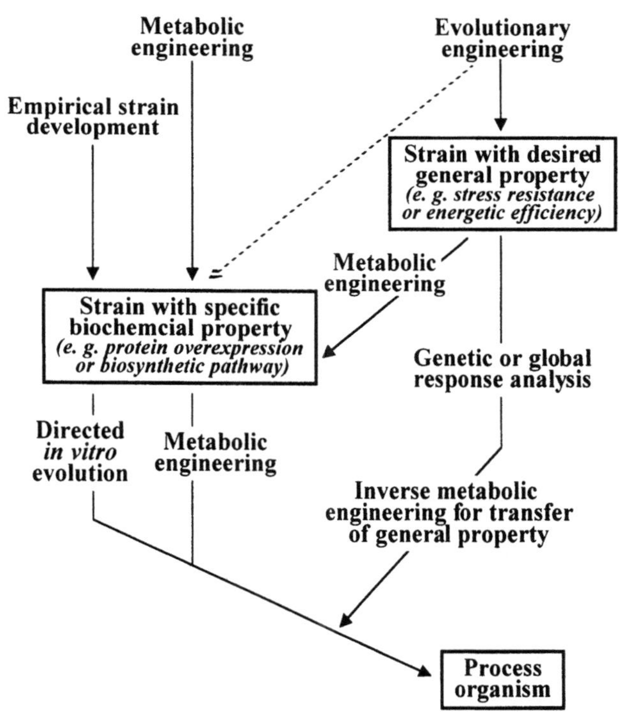

Fig. 7. Flow chart for future biotechnological strain development. The *dashed arrow* indicates a less likely but possible route

may have been the primary reason why, with few exceptions [134, 139], this road has remained almost untrodden in biotechnological research. However, recent technological advances are rapidly changing this situation and inverse metabolic engineering is likely to gain more relevance in the near future. Mass sequencing and functional genomics are currently the most effective approaches for increasing such knowledge at the molecular level of different organisms. Several methods that provide access to global cellular responses can now routinely be used for the identification of the molecular bases for useful phenotypes. One example is simultaneous and comprehensive analysis of gene expression at the protein level by two-dimensional protein gel electrophoresis in combination with genomic sequence information and mass spectrometric spot identification. This is often referred to as *proteome* analysis [142]. Similarly, genome-wide mRNA levels can be monitored by so-called *transcriptome* analysis, which is based upon extraction of total mRNA that is then hybridized to arrays of oligonucleotides or open reading frames arranged on DNA chips or membranes [143]. Successful identification of the molecular basis for evolved phenotypes through these technologies includes proteome analysis of *E. coli* variants [138, 144] and transcriptome analysis of improved yeast variants [134].

An alternative application of DNA chips in evolutionary engineering is the rapid identification of beneficial or detrimental genes with respect to a particular phenotype in selection experiments. Briefly, hybridizing PCR-amplified DNA from positively selected clones to a genomic DNA chip of this organism can reveal enrichment or depletion of clones from an overexpression library as a consequence of a selection procedure [145]. Similar to, but more rapid than, the signature-tagged mutagenesis introduced in Sect. 2.4, this strategy provides access to genes that confer a selective advantage or disadvantage upon overexpression.

Supported by complementary information on global responses at both the metabolite [101] and the flux level [94, 96, 98] (see also Sect. 3.8), these methodologies will pave the road to efficient revelations of the molecular and functional bases of phenotypic variations, even for multifactorial changes. Such global cellular response analyses provide detailed comparative information on many aspects of cellular metabolism, and thus can provide leads to genes that are likely to be involved in a particular phenotype. However, global response analysis cannot directly reveal the mutation(s) that will cause the desired phenotype. Consequently, endowing useful phenotypes on other hosts by inverse metabolic engineering requires intellectual and/or computational interpretation of the results, followed by formulation of hypotheses that would then have to be verified experimentally. Genetic methods that provide more direct access to genomic alterations include genome sequencing, single nucleotide polymorphism, and restriction fragment length polymorphism mapping. Recent developments that make these genetic methods and global response analyses widely available are also expected to stimulate activities in evolutionary engineering.

Acknowledgements. I am most indebted to Jay Bailey for his continuous support and first introducing me to this field. Furthermore, I thank Dan Lasko for critical reading of the manuscript. Our research in evolutionary engineering was supported by the Swiss Priority Program in Biotechnology (SPP BioTech).

References

1. Bailey JE (1991) Science 252:1668
2. Lee SY, Papoutsakis ET (1999) Metabolic engineering. Marcel Dekker
3. Cameron DC, Chaplen FWR (1997) Curr Opin Biotechnol 8:175
4. Bailey JE, Sburlati A, Hatzimanikatis V, Lee K, Renner WA, Tsai PS (1996) Biotechnol Bioeng 52:109
5. Emmerling M, Bailey JE, Sauer U (1999) Metabolic Eng 1:117
6. Rubingh DN (1999) Curr Opin Biotechnol 8:417
7. Marrs B, Delagrave S, Murphy D (1999) Curr Opin Microbiol 2:241
8. Steipe B (2000) Curr Top Microbiol Immunol 243:55
9. Minshull J, Stemmer WPC (1999) Curr Opin Chem Biol 3:284
10. Arnold FH, Volkov A (1999) Curr Opin Chem Biol 3:54
11. Arnold FH (1998) Acc Chem Res 31:125
12. Kettling U, Koltermann A, Eigen M (2000) Curr Top Microbiol Immunol 243:173
13. Stemmer W (2000) Engineering of complex systems by molecular breeding. Metabolic Engineering III Conference, Colorado Springs, CO 22–27 October 2000
14. Georgiou G, DeWitt N (1999) Nat Biotechnol 17:1161
15. Kauffman SA (1993) The origins of order. Oxford University Press, Oxford
16. Wright S (1988) Am Nat 131:115
17. Wright S (1982) Evolution 36:427
18. Kucher O, Arnold FH (1997) Trends Biotechnol 15:523
19. Hall BG (1999) FEMS Microbiol Lett 178:1
20. Parekh S, Vinci VA, Strobel RJ (2000) Appl Microbiol Biotechnol 54:287
21. Rowlands RT (1984) Enz Microbial Technol 6:3
22. Vinci VA, Byng G (1999) Strain improvement by nonrecombinant methods. In: Demain AL, Davies JE (eds.) Manual of industrial microbiology and biotechnology. ASM Press, Washington DC, p 103
23. Müller HJ (1964) Mutat Res 1:2
24. Andersson SGE, Kurland CG (1998) Trends Microbiol 6:263
25. Butler PR, Brown M, Oliver SG (1996) Biotechnol Bioeng 49:185
26. Drake JW (1991) Annu Rev Genet 25:124
27. Moxon ER, Thaler DS (1997) Nature 387:659
28. de Visser JAGM, Zeyl CW, Gerrish PJ, Blanchard JL, Lenski RE (1999) Science 283:404
29. Friedberg EC, Walker GC, Siede W (1995) DNA repair and mutagenesis. ASM Press, Washington, DC
30. Arber W (2000) FEMS Microbiol Rev 24:1
31. Radman M (1999) Nature 401:866
32. MacPhee DG (1993) ASM Press 59:297
33. Liu LX, Spoerke JM, Mulligan EL, Chen J, Reardon B, Westlund B, Sun L, Abel K, Armstrong B, Hardiman G, King J, McCague L, Basson M, Clover R, Johnson CD (1999) Genome Res 9:859
34. Rowlands RT (1983) Industrial fungal genetics and strain selection. In: Smith JE, Berry DR, Kristiansen B (eds.) The filamentous fungi, vol 4. Edward Arnold, London, p 346
35. Kaplan NL, Hudson RR, Langley CH (1989) Genetics 123:887
36. Harder W, Kuenen JG, Matin A (1977) J Appl Bacteriol 43:1
37. Sniegowski PD, Gerrish PJ, Lenski RE (1997) Nature 387:703
38. McBeth DL, Hauer B (1996) Appl Environ Microbiol 62:3538
39. Taddei F, Radman M, Maynard-Smith J, Toupance B, Gouyon PH, Godelle B (1997) Nature 387:700
40. Liao H, McKenzie T, Hageman R (1986) Proc Natl Acad Sci USA 83:576
41. Low NM, Holliger P, Winter G (1996) J Mol Biol 260:359
42. Long-McGie J, Liu AD, Schellenberger V (2000) Biotechnol Bioeng 68:121
43. Miller JH (1996) Annu Rev Microbiol 50:625

44. Xiao W, Chow BL, Fontanie T, Ma L, Bacchetti S, Hryciw T, Broomfield S (1999) Mutat Res 435:1
45. Chen C, Merrill BJ, Lau PJ, Holm C, Kolodner RD (1999) Mol Cell Biol 19:7801
46. Miller JH, Suthar A, Tai J, Yeung A, Truong C, Stewart JL (1999) J Bacteriol 181:1576
47. de Lorenzo V, Herrero M, Sanchez JM, Timmis KN (1998) FEMS Microbiol Ecol 27:211
48. Snyder L, Champness W (1997) Molecular genetics of bacteria. ASM Press, Washington, DC
49. Hensel M (1998) Electrophoresis 19:608
50. Shoemaker DD, Lashkari DA, Morris D, Mittmann M, Davis RW (1996) Nat Genet 14:450
51. Lee MS, Dougherty BA, Madeo AC, Morrison DA (1999) Appl Environ Microbiol 65:1883
52. Smith JM, Smith NH, O'Rourke M, Spratt BG (1993) Proc Natl Acad Sci USA 90:4384
53. Paquin C, Adams J (1983) Nature 302:495
54. Orr HA, Otto SP (1994) Genetics 136:1475
55. Seegers JFML, Franke CM, Kiewiet R, Venema G, Bron S (1995) Plasmid 33:71
56. Velicer GJ (1999) Appl Environ Microbiol 65:264
57. Weikert C, Sauer U, Bailey JE (1997) Microbiol 143:1567
58. Çakar ZP, Sonderegger M, Sauer U (2001) (Submitted for publication)
59. Hall BG, Yokohama S, Calhoun DH (1983) Mol Biol Evol 1:109
60. Berg OG (1995) J Theor Biol 173:307
61. Dykhuizen DE (1990) Annu Rev Ecol Syst 21:373
62. Eberhard A (1972) J Bacteriol 109:101
63. Bierbaum G, Karutz M, Weuster-Botz D, Wandrey C (1994) Appl Microbiol Biotechnol 40:611
64. Walker JE, Miroux B (1999) Selection of *Escherichia coli* hosts that are optimized for the expression of proteins. In: Demain AL, Davies JE (eds) Manual of industrial microbiology and biotechnology. ASM Press, Washington, DC, p 575
65. Papadopoulos D, Schneider D, Meier-Eiss J, Arber W, Lenski RE, Blot M (1999) Proc Natl Acad Sci USA 96:3807
66. Lenski RE, Mongold JA, Sniegowski PD, Travisano M, Vasi F, Gerrish PJ, Schmidt TM (1998) Ant Leeuwenhoek 73:35
67. Lenski RE, Travisano M (1994) Proc Natl Acad Sci USA 91:6808
68. Naki D, Paech C, Ganshaw G, Schellenberger V (1998) Appl Microbiol Biotechnol 49: 290
69. Dykhuizen DE, Hartl DL (1983) Microbiol Rev 47:150
70. Dykhuizen DE (1993) Meth Enz 224:613
71. Teixeira de Mattos MJ, Neijssel OM (1997) J Biotechnol 59:117
72. Dawson PSS (1985) CRC Crit Rev Biotechnol 2:315
73. Kurland CG (1992) Annu Rev Genet 26:29
74. Helling RB, Vargas CN, Adams J (1987) Genetics 116:349
75. Rosenzweig RF, Sharp RR, Treves DS, Adams J (1994) Genetics 137:1
76. Treves DS, Manning S, Adams J (1998) Mol Biol Evol 15:789
77. Paquin CE, Adams J (1983) Nature 306:368
78. Gostomski P, Mühlemann M, Lin Y-H, Mormino R, Bungay H (1994) J Biotechnol 37:167
79. Zeng A-P (1999) Continuous culture. In: Demain AL, Davies JE (eds) Manual of industrial microbiology and biotechnology. ASM Press, Washington, DC, p 151
80. Bryson V, Szybalski W (1952) Science 116:45
81. Fleming G, Dawson MT, Patching JW (1988) J Gen Microbiol 134:2095
82. Brown SW, Oliver SG (1982) Eur J Appl Microbiol Biotechnol 16:119
83. Lane PG, Hutter A, Oliver SG, Butler PR (1999) Biotechnol Prog 15:1115
84. Lane PG, Oliver SG, Butler PR (1999) Biotechnol Bioeng 65:397
85. Orr HA (1999) Genet Res Camb 74:207
86. Kimura M (1983) The neutral theory of molecular evolution. Cambridge University Press, Cambridge, UK
87. Orr HA (1998) Evolution 52:935
88. Davey HM, Jones A, Shaw AD, Kell DB (1999) Cytometry 35:162

89. Koch AL (1994) Growth measurement. In: Gerhardt P, Murray GGE, Wood WA, Krieg NR (eds) Methods for general and molecular bacteriology. ASM Press, Washington, DC, p 248
90. Çakar ZP (2000) Diss. ETH No 13665. ETH Zürich
91. Goodacre R, Trew S, Wrigley-Jones C, Neal M, Maddock J, Ottley TW, Porter N, Kell DB (1994) Biotechnol Bioeng 44:1205
92. Lasko DR, Zamboni N, Sauer U (2000) Appl Microbiol Biotechnol 54:243
93. Duetz WA, Rüedi L, Hermann R, O'Connor K, Büchs J, Witholt B (2000) Appl Environ Microbiol 66:2641
94. Varma A, Palsson BO (1994) Bio/Technol 12:994
95. Szyperski T (1998) Q Rev Biophys 31:41
96. Sauer U, Szyperski T, Bailey JE (2000) Future trends in complex microbial reaction studies. In: Barbotin J-N, Portais J-C (eds) NMR in microbiology: theory and applications. Horizon Scientific Press, Wymondham, UK, p 479
97. Szyperski T (1995) Eur J Biochem 232:433
98. Sauer U, Lasko DR, Fiaux JMH, Glaser R, Szyperski T, Wüthrich K, Bailey JE (1999) J Bacteriol 181:6679
99. Christensen B, Nielsen J (1999) Metabolic Eng 1:282
100. Dauner M, Sauer U (2000) Biotechnol Progr 16:642
101. Oliver SG, Winson MK, Kell DB, Baganz F (1998) Trends Biotechnol 16:373
102. Varner J, Ramkrishna D (1999) Curr Opin Biotechnol 10:146
103. Bailey JE (1998) Biotechnol Prog 14:8
104. Dykhuizen DE, Dean AM (1990) Trends Environ Ecol 5:257
105. Tsen S-D, Lai S-C, Pang C-P, Lee J-I, Wilson TH (1996) Biochem Biophys Res Comm 224:351
106. Flores N, Xiao J, Berry A, Bolivar F, Valle F (1996) Nature Biotechnol 14:620
107. Chen R, Hatzimanikatis V, Yap WM, Postma PW, Bailey JE (1997) Biotechnol Prog 13:768
108. Ragout A, Sineriz F, Kaul R, Guoqiang D, Mattisson B (1996) Appl Microbiol Biotechnol 46:126
109. Cornish A, Greenwood JA, Jones CW (1989) J Gen Microbiol 135:3001
110. Silman NJ, Carver MA, Jones CW (1989) J Gen Microbiol 135:3153
111. Mortlock RP, Gallo MA (1992) Experiments in the evolution of catabolic pathways using modern bacteria. In: Mortlock RP (eds) The evolution of metabolic function. CRC Press, Boca Raton, p 1
112. Zelder O, Hauer B (2000) Curr Opin Microbiol 3:248
113. van der Meer JR (1997) Ant Leeuwenhoek 71:159
114. Hall BG, Hauer B (1993) Meth Enz 224:603
115. Schneider K-H, Jäkel G, Hoffmann R, Giffhorn F (1995) Microbiol 141:1865
116. Torkelson J, Harris RS, Lombardo M-J, Nadgendran CT, Rosenberg S (1997) EMBO J 16:3303
117. Finkel SE, Kolter R (1999) Proc Natl Acad Sci USA 96:4023
118. Aarnio TH, Suihko M-L, Kauppinen VS (1991) Appl Biochem Biotechnol 27:55
119. Schellenberger V (2000) Biotechnology 2000, Berlin, Germany, 3–8 September 2000
120. Ebner H, Follmann H (1983) Acetic acid. In: Rehm H-J, Reed G (eds) Biomass, microorganisms for special applications, microbial products I, energy from renewable resources, vol 3. Verlag Chemie, Weinheim, Germany, p 387
121. Tsen S-D (1994) Appl Microbiol Biotechnol 41:233
122. Sauer U, Hatzimanikatis V, Hohmann H-P, Manneberg M, van Loon APGM, Bailey JE (1996) Appl Environ Microbiol 62:3687
123. Kurland CG, Dong H (1996) Mol Microbiol 21:1
124. Miroux B, Walker JE (1996) J Mol Biol 260:289
125. Noack D, Geuther R, Tonew M, Breitling R, Behnke D (1988) Gene 68:53
126. O'Kennedy RD, Patching JW (1999) J Biotechnol 69:203
127. Hjortso MA, Bailey JE (1984) Biotechnol Bioeng 26:528
128. Lenski RE, Simpson SC, Nguyen TT (1994) J Bacteriol 176:3140
129. Wiebe MG, Robson GD, Oliver SG, Trinci APJ (1994) Microbiol 140:3015

130. Wiebe MG, Robson GD, Oliver SG, Trinci APJ (1996) Biotechnol Bioeng 51:61
131. Withers JM, Wiebe MG, Robson GD, Osborne D, Turner G, Trinci APJ (1995) FEMS Microbiol Lett 133:245
132. Withers JM, Wiebe MG, Robson GD, Trinci APJ (1994) Mycol Res 98:95
133. Brown CJ, Todd KM, Rosenzweig RF (1998) Mol Biol Evol 15:931
134. Ferea TL, Botstein D, Brown PO, Rosenzweig RF (1999) Proc Natl Acad Sci USA 96:9721
135. Adams J, Paquin C, Oeller PW, Lee LW (1985) Genetics 110:173
136. Weikert C, Sauer U, Bailey JE (1998) Biotechnol Prog 14:420
137. Weikert C, Sauer U, Bailey JE (1998) Biotechnol Bioeng 59:386
138. Weikert C(1998) Diss. ETH No 12594, ETH Zürich
139. Weikert C, Canonaco F, Sauer U, Bailey JE (2000) Metabolic Eng 2:293–299
140. Huisman GW, Kolter R (1994) Science 265:537
141. Cantor CR (2000) Trends Biotechnol 18:6
142. James P (1997) Q Rev Biophys 30:279
143. Gerhold D, Rushmore T, Caskey CT (1999) Trends Biochem Sci 24:168
144. Kurlandzka A, Rosenzweig RF, Adams J (1991) Mol Biol Evol 8:261
145. Kao CM (1999) Biotechnol Prog 15:304
146. Zhao H, Moore JC, Volkov AA, Arnold FH (1999) Methods for optimizing industrial enzymes by directed evolution. In: Demain AL, Davies JE (eds) Manual of industrial microbiology and biotechnology. ASM Press, Washington, DC, p 597

Received: November 2000

Author Index Volume 1–73

Author Index Volumes 1–50 see Volume 50

Ackermann, J.-U. see Babel, W.: Vol. 71, p. 125

Adam, W., Lazarus, M., Saha-Möller, C. R., Weichhold, O., Hoch, U. Häring, D., Schreier, Ü.: Biotransformations with Peroxidases. Vol. 63, p. 73

Allan, J. V., Roberts, S. M., Williamson, N. M.: Polyamino Acids as Man-Made Catalysts. Vol. 63, p. 125

Al-Rubeai, M.: Apoptosis and Cell Culture Technology. Vol. 59, p. 225

Al-Rubeai, M. see Singh, R. P.: Vol. 62, p. 167

Alsberg, B. K. see Shaw, A. D.: Vol. 66, p. 83

Antranikian, G. see Ladenstein, R.: Vol. 61, p. 37

Antranikian, G. see Müller, R.: Vol. 61, p. 155

Archelas, A. see Orru, R. V. A.: Vol. 63, p. 145

Argyropoulos, D. S.: Lignin. Vol. 57, p. 127

Arnold, F. H., Moore, J. C.: Optimizing Industrial Enzymes by Directed Evolution. Vol. 58, p. 1

Akhtar, M., Blanchette, R. A., Kirk, T. K.: Fungal Delignification and Biochemical Pulping of Wood. Vol. 57, p. 159

Autuori, F., Farrace, M. G., Oliverio, S., Piredda, L., Piacentini, G.: "Tissie" Transglutaminase and Apoptosis. Vol. 62, p. 129

Azerad, R.: Microbial Models for Drug Metabolism. Vol. 63, p. 169

Babel, W., Ackermann, J.-U., Breuer, U.: Physiology, Regulation and Limits of the Synthesis of Poly(3HB). Vol. 71, p. 125

Bajpai, P., Bajpai, P. K.: Realities and Trends in Emzymatic Prebleaching of Kraft Pulp. Vol. 56, p. 1

Bajpai, P. Bajpai, P. K.: Reduction of Organochlorine Compounds in Bleach Plant Effluents. Vol. 57, p. 213

Bajpai, P. K. see Bajpai, P.: Vol. 56, p. 1

Bajpai, P. K. see Bajpai, P.: Vol. 57, p. 213

Bárzana, E.: Gas Phase Biosensors. Vol. 53, p. 1

Bazin, M. J. see Markov, S. A.: Vol. 52, p. 59

Bellgardt, K.-H.: Process Models for Production of β-Lactam Antibiotics. Vol. 60, p. 153

Beppu, T.: Development of Applied Microbiology to Modern Biotechnology in Japan. Vol. 69, p. 41

Berovic, M. see, Mitchell, D. A.: Vol. 68, 61

Beyeler, W., DaPra, E., Schneider, K.: Automation of Industrial Bioprocesses. Vol. 70, p. 139

Beyer, M. see Seidel, G.: Vol. 66, p. 115

Bhatia, P. K., Mukhopadhyay, A.: Protein Glycosylation: Implications for in vivo Functions and Thereapeutic Applications. Vol. 64, p. 155

Bisaria, V. S. see Ghose, T. K.: Vol. 69, p. 87

Blanchette R. A. see Akhtar, M.: Vol. 57, p. 159

de Bont, J. A. M. see van der Werf, M. J.: Vol. 55, p. 147

Bocker, H., Knorre, W. A.: Antibiotica Research in Jena from Penicillin and Nourseothricin to Interferon. Vol. 70, p. 35

Brainard, A. P. see Ho, N. W. Y.: Vol. 65, p. 163
Breuer, U. see Babel, W.: Vol. 71, p. 125
Broadhurst, D. see Shaw, A. D.: Vol. 66, p. 83
Bruckheimer, E. M., Cho, S. H., Sarkiss, M., Herrmann, J., McDonell, T. J.: The Bcl-2 Gene Family and Apoptosis. Vol 62, p. 75
Buchert, J. see Suurnäkki, A.: Vol. 57, p. 261
Bungay, H. R. see Mühlemann, H. M.: Vol. 65, p. 193
Bungay, H. R., Isermann, H. P.: Computer Applications in Bioprocessin. Vol. 70, p. 109
Byun, S. Y. see Choi, J. W.: Vol. 72, p. 63

Cao, N. J. see Gong, C. S.: Vol. 65, p. 207
Cao, N. J. see Tsao, G. T.: Vol. 65, p. 243
Carnell, A. J.: Stereoinversions Using Microbial Redox-Reactions. Vol. 63, p. 57
Cen, P., Xia, L.: Production of Cellulase by Solid-State Fermentation. Vol. 65, p. 69
Chang, H. N. see Lee, S. Y.: Vol. 52, p. 27
Cheetham, P. S. J.: Combining the Technical Push and the Business Pull for Natural Flavours. Vol. 55, p. 1
Cho, S. H. see Bruckheimer, E. M.: Vol. 62, p. 75
Cho, G. H. see Choi, J. W.: Vol 72, p. 63
Choi, J. see Lee, S. Y.: Vol. 71, p. 183
Choi, J. W., Cho, G. H., Byun, S. Y., Kim, D.-I.: Integrated Bioprocessing for Plant Cultures. Vol. 72, p. 63
Chen, Z. see Ho, N. W. Y.: Vol. 65, p. 163–192
Christensen, B., Nielsen, J.: Metabolic Network Analysis – A Powerful Tool in Metabolic Engineering. Vol. 66, p. 209
Ciaramella, M. see van der Oost, J.: Vol. 61, p. 87
Contreras, B. see Sablon, E.: Vol. 68, p. 21
Cordero Otero, R. R. see Hahn-Hägerdal, B.: Vol. 73, p. 53
Cornet, J.-F., Dussap, C. G., Gros, J.-B.: Kinetics and Energetics of Photosynthetic Micro-Organisms in Photobioreactors. Vol. 59, p. 153
da Costa, M. S., Santos, H., Galinski, E. A.: An Overview of the Role and Diversity of Compatible Solutes in Bacteria and Archaea. Vol. 61, p. 117
Cotter, T. G. see McKenna, S. L.: Vol. 62, p. 1
Croteau, R. see McCaskill, D.: Vol. 55, p. 107

Danielsson, B. see Xie, B.: Vol. 64, p. 1
DaPra, E. see Beyeler, W.: Vol. 70, p. 139
Darzynkiewicz, Z., Traganos, F.: Measurement of Apoptosis. Vol 62, p. 33
Davey, H. M. see Shaw, A. D.: Vol. 66, p. 83
Dean, J. F. D., LaFayette, P. R., Eriksson, K.-E. L., Merkle, S. A.: Forest Tree Biotechnolgy. Vol. 57, p. 1
Demain, A. L., Fang, A.: The natural Functions of Secondary Metabolites. Vol. 69, p. 1
Dochain, D., Perrier, M.: Dynamical Modelling, Analysis, Monitoring and Control Design for Nonlinear Bioprocesses. Vol. 56, p. 147
Du, J. see Gong, C. S: Vol. 65, p. 207–241
Du, J. see Tsao, G. T.: Vol. 65, p. 243–280
Dueser, M. see Raghavarao, K. S. M. S.: Vol. 68, p. 139
Dussap, C. G. see Cornet J.-F.: Vol. 59, p. 153
Dutta, N. N. see Ghosh, A. C.: Vol. 56, p. 111
Dynesen, J. see McIntyre, M.: Vol. 73, p. 103

Eggeling, L., Sahm, H., de Graaf, A. A.: Quantifying and Directing Metabolite Flux: Application to Amino Acid Overproduction. Vol. 54, p. 1
Eggeling, L. see de Graaf, A. A.: Vol. 73, p. 9
Eggink, G., see Kessler, B.: Vol. 71, p. 159

Eggink, G., see van der Walle, G. J. M.: Vol. 71, p. 263
Ehrlich, H. L. see Rusin, P.: Vol. 52, p. 1
Elias, C. B., Joshi, J. B.: Role of Hydrodynamic Shear on Activity and Structure of Proteins. Vol. 59, p. 47
Elling, L.: Glycobiotechnology: Enzymes for the Synthesis of Nucleotide Sugars. Vol. 58, p. 89
Eriksson, K.-E. L. see Kuhad, R. C.: Vol. 57, p. 45
Eriksson, K.-E. L. see Dean, J. F. D.: Vol. 57, p. 1

Faber, K. see Orru, R. V. A.: Vol. 63, p. 145
Fang, A. see Demain, A. L.: Vol. 69, p. 1
Farrell, R. L., Hata, K., Wall, M. B.: Solving Pitch Problems in Pulp and Paper Processes. Vol. 57, p. 197
Farrace, M. G. see Autuori, F.: Vol. 62, p. 129
Fiechter, A.: Biotechnology in Switzerland and a Glance at Germany. Vol. 69, p. 175
Fiechter, A. see Ochsner, U. A.: Vol. 53, p. 89
Foody, B. see Tolan, J. S.: Vol. 65, p. 41
Freitag, R., Hórvath, C.: Chromatography in the Downstream Processing of Biotechnological Products. Vol. 53, p. 17
Furstoss, R. see Orru, R. V. A.: Vol. 63, p. 145

Galinski, E. A. see da Costa, M. S.: Vol. 61, p. 117
Gàrdonyi, M. see Hahn-Hägerdal, B.: Vol. 73, p. 53
Gatfield, I. L.: Biotechnological Production of Flavour-Active Lactones. Vol. 55, p. 221
Gemeiner, P. see Stefuca, V.: Vol. 64, p. 69
Gerlach, S. R. see Schügerl, K.: Vol. 60, p. 195
Ghose, T. K., Bisaria, V. S.: Development of Biotechnology in India. Vol. 69, p. 71
Ghosh, A. C., Mathur, R. K., Dutta, N. N.: Extraction and Purification of Cephalosporin Antibiotics. Vol. 56, p. 111
Ghosh, P. see Singh, A.: Vol. 51, p. 47
Gilbert, R. J. see Shaw, A. D.: Vol. 66, p. 83
Gill, R. T. see Stephanopoulos, G.: Vol. 73, p. 1
Gomes, J., Menawat, A. S.: Fed-Batch Bioproduction of Spectinomycin. Vol. 59, p. 1
Gong, C. S., Cao, N. J., Du, J., Tsao, G. T.: Ethanol Production from Renewable Resources. Vol. 65, p. 207–241
Gong, C. S. see Tsao, G. T.: Vol. 65, p. 243
Goodacre, R. see Shaw, A. D.: Vol. 66, p. 83
de Graaf, A. A., Eggeling, L., Sahm, H.: Metabolic Engineering for L-Lysine Production by *Corynebacterium glutamicum*. Vol. 73, p. 9
de Graaf, A.A. see Eggeling, L.: Vol. 54, p. 1
de Graaf, A.A. see Weuster-Botz, D.: Vol. 54, p. 75
de Graaf, A.A. see Wiechert, W.: Vol. 54, p. 109
Grabley, S., Thiericke, R.: Bioactive Agents from Natural Sources: Trends in Discovery and Application. Vol. 64, p. 101
Griengl, H. see Johnson, D. V.: Vol. 63, p. 31
Gros, J.-B. see Larroche, C.: Vol. 55, p. 179
Gros, J.-B. see Cornet, J. F.: Vol. 59, p. 153
Guenette M. see Tolan, J. S.: Vol. 57, p. 289
Gutman, A. L., Shapira, M.: Synthetic Applications of Enzymatic Reactions in Organic Solvents. Vol. 52, p. 87

Häring, D. see Adam, E.: Vol. 63, p. 73
Hahn-Hägerdal, B., Wahlbom, Gárdonyi, M., van Zyl, W. H., Cordero Otero, R. R., Jönsson, L. J.: Metabolic Engineering of *Saccharomyces cerevisiae* for Xylose Utilization. Vol. 73, p. 53
Haigh, J. R. see Linden, J. C.: Vol. 72, p. 27
Hall, D. O. see Markov, S. A.: Vol. 52, p. 59

Hall, P. see Mosier, N. S.: Vol. 65, p. 23
Harvey, N. L., Kumar, S.: The Role of Caspases in Apoptosis. Vol. 62, p. 107
Hasegawa, S., Shimizu, K.: Noninferior Periodic Operation of Bioreactor Systems. Vol. 51, p. 91
Hata, K. see Farrell, R. L.: Vol. 57, p. 197
van der Heijden, R. see Memelink, J.: Vol. 72, p. 103
Hein, S. see Steinbüchel, A.: Vol. 71, p. 81
Hembach, T. see Ochsner, U. A.: Vol. 53, p. 89
Henzler, H.-J.: Particle Stress in Bioreactor. Vol. 67, p. 35
Herrmann, J. see Bruckheimer, E. M.: Vol. 62, p. 75
Hill, D. C., Wrigley, S. K., Nisbet, L. J.: Novel Screen Methodologies for Identification of New Microbial Metabolites with Pharmacological Activity. Vol. 59, p. 73
Hiroto, M. see Inada, Y.: Vol. 52, p. 129
Ho, N. W. Y., Chen, Z., Brainard, A. P. Sedlak, M.: Successful Design and Development of Genetically Engineering Saccharomyces Yeasts for Effective Cofermentation of Glucose and Xylose from Cellulosic Biomass to Fuel Ethanol. Vol. 65, p. 163
Hoch, U. see Adam, W.: Vol. 63, p. 73
Holló, J., Kralovánsky, U. P.: Biotechnology in Hungary. Vol. 69, p. 151
Honda, H., Liu, C., Kobayashi, T.: Large-Scale Plant Micropropagation. Vol. 72, p. 157
Hórvath, C. see Freitag, R.: Vol. 53, p. 17
Hummel, W.: New Alcohol Dehydrogenases for the Synthesis of Chiral Compounds. Vol. 58, p.145

Inada, Y., Matsushima, A., Hiroto, M., Nishimura, H., Kodera, Y.: Chemical Modifications of Proteins with Polyethylen Glycols. Vol. 52, p. 129
Iyer, P. see Lee, Y. Y.: Vol. 65, p. 93–115
Irwin, D. C. see Wilson, D. B.: Vol. 65, p. 1
Isermann, H. P. see Bungay, H. R.: Vol. 70, p. 109

James, E., Lee, J. M.: The Production of Foreign Proteins from Genetically Modified Plant Cells. Vol. 72, p. 127
Jeffries, T. W., Shi, N.-Q.: Genetic Engineering for Improved Xylose Fementation by Yeasts. Vol. 65, p. 117
Jendrossek, D.: Microbial Degradation of Polyesters. Vol. 71, p. 293
Jönsson, L. J. see Hahn-Hägerdal, B.: Vol. 73, p. 53
Johnson, E. A., Schroeder, W. A.: Microbial Carotenoids. Vol. 53, p. 119
Johnson, D. V., Griengl, H.: Biocatalytic Applications of Hydroxynitrile. Vol. 63, p. 31
Joshi, J. B. see Elias, C. B.: Vol. 59, p. 47
Johnsurd, S. C.: Biotechnolgy for Solving Slime Problems in the Pulp and Paper Industry. Vol. 57, p. 311

Kaderbhai, N. see Shaw, A. D.: Vol. 66, p. 83
Kataoka, M. see Shimizu, S.: Vol. 58, p. 45
Kataoka, M. see Shimizu, S.: Vol. 63, p. 109
Katzen, R., Tsao, G. T.: A View of the History of Biochemical Engineering. Vol. 70, p. 77
Kawai, F.: Breakdown of Plastics and Polymers by Microorganisms. Vol. 52, p. 151
Kell, D. B. see Shaw, A. D.: Vol. 66, p. 83
Kessler, B., Weusthuis, R., Witholt, B., Eggink, G.: Production of Microbial Polyesters: Fermentation and Downstream Processes. Vol. 71, p. 159
Khosla, C. see McDaniel, R.: Vol. 73, p. 31
Kieran, P. M., Malone, D. M., MacLoughlin, P. F.: Effects of Hydrodynamic and Interfacial Forces on Plant Cell Suspension Systems. Vol. 67, p. 139
Kijne, J. W. see Memelink, J.: Vol. 72, p. 103
Kim, D.-I. see Choi, J. W.: Vol. 72, p. 63
Kim, Y. B., Lenz, R. W.: Polyesters from Microorganisms. Vol. 71, p. 51

King, R.: Mathematical Modelling of the Morphology of Streptomyces Species. Vol. 60, p. 95
Kino-oka, M., Nagatome, H., Taya, M.: Characterization and Application of Plant Hairy Roots Endowed with Photosynthetic Functions. Vol. 72, p. 183
Kirk, T. K. see Akhtar, M.: Vol. 57, p. 159
Knorre, W. A. see Bocker, H.: Vol. 70, p. 35
Kobayashi, M. see Shimizu, S.: Vol. 58, p. 45
Kobayashi, T. see Honda, H.: Vol. 72, p. 157
Kobayashi, S., Uyama, H.: *In vitro* Biosynthesis of Polyesters. Vol. 71, p. 241
Kodera, F. see Inada, Y.: Vol. 52, p. 129
Kolattukudy, P. E.: Polyesters in Higher Plants. Vol. 71, p. 1
de Koning, G. J. M. see van der Walle, G. A. M.: Vol. 71, p. 263
Kossen, N. W. F.: The Morphology of Filamentous Fungi. Vol. 70, p. 1
Krabben, P. Nielsen, J.: Modeling the Mycelium Morphology of Penicilium Species in Submerged Cultures. Vol. 60, p. 125
Kralovánszky, U. P. see Holló, J.: Vol. 69, p. 151
Krämer, R.: Analysis and Modeling of Substrate Uptake and Product Release by Procaryotic and Eucaryotik Cells. Vol. 54, p. 31
Kretzmer G.: Influence of Stress on Adherent Cells. Vol. 67, p. 123
Krieger, N. see Mitchell, D. A.: Vol. 68, p. 61
Kuhad, R. C., Singh, A., Eriksson, K.-E. L.: Microorganisms and Enzymes Involved in the Degradation of Plant Cell Walls. Vol. 57, p. 45
Kuhad, R. Ch. see Singh, A.: Vol. 51, p. 47
Kumagai, H.: Microbial Production of Amino Acids in Japan. Vol. 69, p. 71
Kumar, S. see Harvey, N. L.: Vol. 62, p. 107

Ladenstein, R., Antranikian, G.: Proteins from Hyperthermophiles: Stability and Enzamatic Catalysis Close to the Boiling Point of Water. Vol. 61, p. 37
Ladisch, C. M. see Mosier, N. S.: Vol. 65, p. 23
Ladisch, R. M. see Mosier, N. S.: Vol. 65, p. 23
Lammers, F., Scheper, T.: Thermal Biosensors in Biotechnology. Vol. 64, p. 35
Larroche, C., Gros, J.-B.: Special Transformation Processes Using Fungal Spares and Immobilized Cells. Vol. 55, p. 179
LaFayette, P. R. see Dean, J. F. D.: Vol. 57, p. 1
Lazarus, M. see Adam, W.: Vol. 63, p. 73
Leak, D. J. see van der Werf, M. J.: Vol. 55, p. 147
Lee, S. Y., Chang, H. N.: Production of Poly(hydroxyalkanoic Acid). Vol. 52, p. 27
Lee, S. Y., Choi, J.: Production of Microbial Polyester by Fermentation of Recombinant Microorganisms. Vol. 71, p. 183
Lee, Y. Y., Iyer, P., Torget, R. W.: Dilute-Acid Hydrolysis of Lignocellulosic Biomass. Vol. 65, p. 93
Lee, J. M. see James, E.: Vol. 72, p. 127
Lenz, R. W. see Kim, Y. B.: Vol. 71. p. 51
Licari, P. see McDaniel, R.: Vol. 73, p. 31
Lievense, L. C., van't Riet, K.: Convective Drying of Bacteria II. Factors Influencing Survival. Vol. 51, p. 71
Linden, J. C., Haigh, J. R., Mirjalili, N., Phisaphalong, M.: Gas Concentration Effects on Secondary Metabolite Production by Plant Cell Cultures. Vol. 72, p. 27
Liu, C. see Honda, H.: Vol. 72, p. 157

MacLoughlin, P. F. see Kieran, P. M.: Vol. 67, p. 139
Malone, D. M. see Kieran, P. M.: Vol. 67, p. 139
Maloney, S. see Müller, R.: Vol. 61, p. 155
Mandenius, C.-F.: Electronic Noses for Bioreactor Monitoring. Vol. 66, p. 65
Markov, S. A., Bazin, M. J., Hall, D. O.: The Potential of Using Cyanobacteria in Photobioreactors for Hydrogen Production. Vol. 52, p. 59
Marteinsson, V. T. see Prieur, D.: Vol. 61, p. 23

Mathur, R. K. see Ghosh, A. C.: Vol. 56, p. 111
Matsushima, A. see Inada, Y.: Vol. 52, p. 129
McCaskill, D., Croteau, R.: Prospects for the Bioengineering of Isoprenoid Biosynthesis. Vol. 55, p. 107
McDaniel, R., Licari, P., Khosla, C.: Process Development and Metabolic Engineering for the Overproduction of Natural and Unnatural Polyketides. Vol. 73, p. 31
McDonell, T. J. see Bruckheimer, E. M.: Vol. 62, p. 75
McGovern, A. see Shaw, A. D.: Vol. 66, p. 83
McGowan, A. J. see McKenna, S. L.: Vol. 62, p. 1
McIntyre, M., Müller, C., Dynesen, J., Nielsen, J.: Metabolic Engineering of the *Aspergillus*. Vol. 73, p. 103
McKenna, S. L.: McGowan, A. J., Cotter, T. G.: Molecular Mechanisms of Programmed Cell Death. Vol. 62, p. 1
McLoughlin, A. J.: Controlled Release of Immobilized Cells as a Strategy to Regulate Ecological Competence of Inocula. Vol. 51, p. 1
Memelink, J., Kijne, J. W., van der Heijden, R., Verpoorte, R.: Genetic Modification of Plant Secondary Metabolite Pathways Using Transcriptional Regulators. Vol. 72, p. 103
Menachem, S. B. see Argyropoulos, D. S. : Vol. 57, p. 127
Menawat, A. S. see Gomes J.: Vol. 59, p. 1
Menge, M. see Mukerjee, J.: Vol. 68, p. 1
Merkle, S. A. see Dean, J. F. D.: Vol. 57, p. 1
Mirjalili, N. see Linden, J. C.: Vol. 72, p. 27
Mitchell, D. A., Berovic, M., Krieger, N.: Biochemical Engineering Aspects of Solid State Bioprocessing. Vol. 68, p. 61
Moore, J. C. see Arnold, F. H.: Vol. 58, p. 1
Mosier, N. S., Hall, P., Ladisch, C. M., Ladisch, M. R.: Reaction Kinetics, Molecular Action, and Mechanisms of Cellulolytic Proteins. Vol. 65, p. 23
Moracci, M. see van der Oost, J.: Vol. 61, p. 87
Mühlemann, H. M. Bungay, H. R.: Research Perspectives for Bioconversion of Scrap Paper. Vol. 65, p. 193
Müller, C. see McIntyre, M.: Vol. 73, p. 103
Müller, R., Antranikian, G., Maloney, S., Sharp, R.: Thermophilic Degradation of Environmental Pollutants. Vol. 61, p. 155
Mukherjee, J., Menge, M.: Progress and Prospects of Ergot Alkaloid Research. Vol. 68, p. 1
Mukhopadhyay, A.: Inclusion Bodies and Purification of Proteins in Biologically Active Forms. Vol. 56, p. 61
Mukhopadhyay, A. see Bhatia, P. K.: Vol. 64, p. 155

Nagatome, H. see Kino-oka, M.: Vol. 72, p. 183
Nielsen, J. see Christensen, B.: Vol. 66, p. 209
Nielsen, J. see Krabben, P.: Vol. 60, p. 125
Nielsen, J. see McIntyre, M.: Vol. 73, p. 103
Nisbet, L. J. see Hill, D. C.: Vol. 59, p. 73
Nishimura, H. see Inada, Y.: Vol. 52, p. 123

Ochsner, U. A., Hembach, T., Fiechter, A.: Produktion of Rhamnolipid Biosurfactants. Vol. 53, p. 89
O'Connor, R.: Survival Factors and Apoptosis: Vol. 62, p. 137
Ogawa, J. see Shimizu, S.: Vol. 58, p. 45
Ohta, H.: Biocatalytic Asymmetric Decarboxylation. Vol. 63, p. 1
van der Oost, J., Ciaramella, M., Moracci, M., Pisani, F. M., Rossi, M., de Vos, W. M.: Molecular Biology of Hyperthermophilic Archaea. Vol. 61, p. 87
Oliverio, S. see Autuori, F.: Vol. 62, p. 129
Orru, R. V. A., Archelas, A., Furstoss, R., Faber, K.: Epoxide Hydrolases and Their Synthetic Applications. Vol. 63, p. 145

Paul, G.C., Thomas, C.R.: Characterisation of Mycelial Morphology Using Image Analysis. Vol. 60, p. 1
Perrier, M. see Dochain, D.: Vol. 56, p. 147
Phisaphalong, M. see Linden, J.C.: Vol. 72, p. 27
Piacentini, G. see Autuori, F.: Vol. 62, p. 129
Piredda, L. see Autuori, F.: Vol. 62, p. 129
Pisani, F.M. see van der Oost, J.: Vol. 61, p. 87
Pohl, M.: Protein Design on Pyruvate Decarboxylase (PDC) by Site-Directed Mutagenesis. Vol. 58, p. 15
Poirier, Y.: Production of Polyesters in Transgenic Plants. Vol. 71, p. 209
Pons, M.-N., Vivier, H.: Beyond Filamentous Species. Vol. 60, p. 61
Pons, M.-N., Vivier, H.: Biomass Quantification by Image Analysis. Vol. 66, p. 133
Prieur, D., Marteinsson, V.T.: Prokaryotes Living Under Elevated Hydrostatic Pressure. Vol. 61, p. 23
Pulz, O., Scheibenbogen, K.: Photobioreactors: Design and Performance with Respect to Light Energy Input. Vol. 59, p. 123

Raghavarao, K.S.M.S., Dueser, M., Todd, P.: Multistage Magnetic and Electrophoretic Extraction of Cells, Particles and Macromolecules. Vol. 68, p. 139
Ramanathan, K. see Xie, B.: Vol. 64, p. 1
van't Riet, K. see Lievense, L. C.: Vol. 51, p. 71
Roberts, S. M. see Allan, J. V.: Vol. 63, p. 125
Roehr, M.: History of Biotechnology in Austria. Vol. 69, p. 125
Rogers, P. L., Shin, H. S., Wang, B.: Biotransformation for L-Ephedrine Production. Vol. 56, p. 33
Rossi, M. see van der Oost, J.: Vol. 61, p. 87
Roychoudhury, P. K., Srivastava, A., Sahai, V.: Extractive Bioconversion of Lactic Acid. Vol. 53, p. 61
Rowland, J. J. see Shaw, A. D.: Vol. 66, p. 83
Rusin, P., Ehrlich, H. L.: Developments in Microbial Leaching – Mechanisms of Manganese Solubilization. Vol. 52, p. 1
Russell, N.J.: Molecular Adaptations in Psychrophilic Bacteria: Potential for Biotechnological Applications. Vol. 61, p. 1

Sablon, E., Contreras, B., Vandamme, E.: Antimicrobial Peptides of Lactic Acid Bacteria: Mode of Action, Genetics and Biosynthesis. Vol. 68, p. 21
Sahai, V. see Singh, A.: Vol. 51, p. 47
Sahai, V. see Roychoudhury, P. K.: Vol. 53, p. 61
Saha-Möller, C. R. see Adam, W.: Vol. 63, p. 73
Sahm, H. see de Graaf, A. A.: Vol. 73, p. 9
Sahm, H. see Eggeling, L.: Vol. 54, p. 1
Saleemuddin, M.: Bioaffinity Based Immobilization of Enzymes. Vol. 64, p. 203
Santos, H. see da Costa, M.S.: Vol. 61, p. 117
Sarkiss, M. see Bruckheimer, E. M.: Vol. 62, p. 75
Sauer, U.: Evolutionary Engineering of Industrially Important Microbial Phenotypes. Vol. 73, p. 129
Scheibenbogen, K. see Pulz, O.: Vol. 59, p. 123
Scheper, T. see Lammers, F.: Vol. 64, p. 35
Schneider, K. see Beyeler, W.: Vol. 70, p. 139
Schreier, P.: Enzymes and Flavour Biotechnology. Vol. 55, p. 51
Schreier, P. see Adam, W.: Vol. 63, p. 73
Schroeder, W. A. see Johnson, E. A.: Vol. 53, p. 119
Schügerl, K., Gerlach, S. R., Siedenberg, D.: Influence of the Process Parameters on the Morphology and Enzyme Production of Aspergilli. Vol. 60, p. 195
Schügerl, K. see Seidel, G.: Vol. 66, p. 115

Schügerl, K.: Recovery of Proteins and Microorganisms from Cultivation Media by Foam Flotation. Vol. 68, p. 191
Schügerl, K.: Development of Bioreaction Engineering. Vol. 70, p. 41
Schumann, W.: Function and Regulation of Temperature-Inducible Bacterial Proteins on the Cellular Metabolism. Vol. 67. p. 1
Schuster, K. C.: Monitoring the Physiological Status in Bioprocesses on the Cellular Level. Vol. 66, p. 185
Scouroumounis, G. K. see Winterhalter, P.: Vol. 55, p. 73
Scragg, A.H.: The Production of Aromas by Plant Cell Cultures. Vol. 55, p. 239
Sedlak, M. see Ho, N. W. Y.: Vol. 65, p. 163
Seidel, G., Tollnick, C., Beyer, M., Schügerl, K.: On-line and Off-line Monitoring of the Production of Cephalosporin C by Acremonium Chrysogenum. Vol. 66, p. 115
Shamlou, P. A. see Yim, S. S.: Vol. 67. p. 83
Shapira, M. see Gutman, A. L.: Vol 52, p. 87
Sharp, R. see Müller, R.: Vol. 61, p. 155
Shaw, A. D., Winson, M. K., Woodward, A. M., McGovern, A., Davey, H. M., Kaderbhai, N., Broadhurst, D., Gilbert, R. J., Taylor, J., Timmins, E. M., Alsberg, B. K., Rowland, J. J., Goodacre, R., Kell, D. B.: Rapid Analysis of High-Dimensional Bioprocesses Using Multivariate Spectroscopies and Advanced Chemometrics. Vol. 66, p. 83
Shi, N.-Q. see Jeffries, T. W.: Vol. 65, p. 117
Shimizu, S., Ogawa, J., Kataoka, M., Kobayashi, M.: Screening of Novel Microbial for the Enzymes Production of Biologically and Chemically Useful Compounds. Vol. 58, p. 45
Shimizu, K. see Hasegawa, S.: Vol. 51, p. 91
Shimizu, S., Kataoka, M.: Production of Chiral C3- and C4-Units by Microbial Enzymes. Vol. 63, p. 109
Shin, H. S. see Rogers, P. L.: Vol. 56, p. 33
Siedenberg, D. see Schügerl, K.: Vol. 60, p. 195
Singh, R. P., Al-Rubeai, M.: Apoptosis and Bioprocess Technology. Vol 62, p. 167
Singh, A., Kuhad, R. Ch., Sahai, V., Ghosh, P.: Evaluation of Biomass. Vol. 51, p. 47
Singh, A. see Kuhad, R. C.: Vol. 57, p. 45
Sonnleitner, B.: New Concepts for Quantitative Bioprocess Research and Development. Vol. 54, p. 155
Sonnleitner, B.: Instrumentation of Biotechnological Processes. Vol. 66, p. 1
Srivastava, A. see Roychoudhury, P. K.: Vol. 53, p. 61
Stafford, D.E., Yanagimachi, K.S., Stephanopoulos, G.: Metabolic Engineering of Indene Bioconversion in *Rhodococcus sp.* Vol. 73, p. 85
Stefuca, V., Gemeiner, P.: Investigation of Catalytic Properties of Immobilized Enzymes and Cells by Flow Microcalorimetry. Vol. 64, p. 69
Steinbüchel, A., Hein, S.: Biochemical and Molecular Basis of Microbial Synthesis of Polyhydroxyalkanoates in Microorganisms. Vol. 71, p. 81
Stephanopoulos, G., Gill, R.T.: After a Decade of Progress, an Expanded Role for Metabolic Engineering. Vol. 73, p. 1
Stephanopoulos, G. see Stafford, D. E.: Vol. 73, p. 85
Suurnäkki, A., Tenkanen, M., Buchert, J., Viikari, L.: Hemicellulases in the Bleaching of Chemical Pulp. Vol. 57, p. 261

Taya, M. see Kino-oka, M.: Vol. 72, p. 183
Taylor, J. see Shaw, A. D.: Vol. 66, p. 83
Tenkanen, M. see Suurnäkki, A.: Vol. 57, p. 261
Thiericke, R. see Grabely, S.: Vol. 64, p. 101
Thömmes, J.: Fluidized Bed Adsorption as a Primary Recovery Step in Protein Purification. Vol. 58, p. 185
Thomas, C. R. see Paul, G. C.: Vol. 60, p. 1
Timmens, E. M. see Shaw, A. D.: Vol. 66, p. 83
Todd, P. see Raghavarao, K. S. M. S.: Vol. 68, p. 139

Tolan, J. S., Guenette, M.: Using Enzymes in Pulp Bleaching: Mill Applications.Vol. 57, p. 289
Tolan, J. S., Foody, B.: Cellulase from Submerged Fermentation. Vol. 65, p. 41
Tollnick, C. see Seidel, G.: Vol. 66, p. 115
Traganos, F. see Darzynkiewicz, Z.: Vol. 62, p. 33
Torget, R. W. see Lee, Y. Y.: Vol. 65, p. 93–115
Tsao, G. T., Cao, N. J. Du, J., Gong, C. S.: Production of Multifunctional Organic Acids from Renewable Resources. Vol. 65, p. 243
Tsao, G. T. see Gong, C. S.: Vol. 65, p. 207
Tsao, G. T. see Katzen, R.: Vol. 70, p. 77

Uyama, H. see Kobayashi, S.: Vol. 71, p. 241

Vandamme, E. see Sablon, E.: Vol. 68, p. 21
Verpoorte, R. see Memelink, J.: Vol. 72, p. 103
Viikari, L. see Suurnäkki, A.: Vol. 57, p. 261
Vivier, H. see Pons, M.-N.: Vol. 60, p. 61
Vivier, H. see Pons, M.-N.: Vol. 66, p. 133
de Vos, W.M. see van der Oost, J.: Vol. 61, p. 87

Wahlbom, C.F. see Hahn-Hägerdal, B.: Vol. 73, p. 53
Wang, B. see Rogers, P. L.: Vol. 56, p. 33
Wall, M. B. see Farrell, R. L.: Vol. 57, p. 197
Weichold, O. see Adam, W.: Vol. 63, p. 73
van der Walle, G. A. M., de Koning, G. J. M., Weusthuis, R. A., Eggink, G.: Properties, Modifications and Applications of Biopolyester. Vol. 71, p. 263
van der Werf, M. J., de Bont, J. A. M. Leak, D. J.: Opportunities in Microbial Biotransformation of Monoterpenes. Vol. 55, p. 147
Weuster-Botz, D., de Graaf, A.A.: Reaction Engineering Methods to Study Intracellular Metabolite Concentrations. Vol. 54, p. 75
Weusthuis, R. see Kessler, B.: Vol. 71, p. 159
Weusthuis, R. A. see van der Walle, G. J. M.: Vol. 71, p. 263
Wiechert, W., de Graaf, A.A.: In Vivo Stationary Flux Analysis by ^{13}C-Labeling Experiments. Vol. 54, p. 109
Wiesmann, U.: Biological Nitrogen Removal from Wastewater. Vol. 51, p. 113
Williamson, N. M. see Allan, J. V.: Vol. 63, p. 125
Wilson, D. B., Irwin, D. C.: Genetics and Properties of Cellulases. Vol. 65, p. 1
Winson, M. K. see Shaw, A. D.: Vol. 66, p. 83
Winterhalter, P., Skouroumounis, G. K.: Glycoconjugated Aroma Compounds: Occurence, Role and Biotechnological Transformation. Vol. 55, p. 73
Witholt, B. see Kessler, B.: Vol. 71, p. 159
Woodley, J.M.: Advances in Enzyme Technology – UK Contributions. Vol. 70, p. 93
Woodward, A. M. see Shaw, A. D.: Vol. 66, p. 83
Wrigley, S. K. see Hill, D. C.: Vol. 59, p. 73

Xia, L. see Cen, P.: Vol. 65, p. 69
Xie, B., Ramanathan, K., Danielsson, B.: Principles of Enzyme Thermistor Systems: Applications to Biomedical and Other Measurements. Vol. 64, p. 1

Yanagimachi, K.S. see Stafford, D.E.: Vol. 73, p. 85
Yim, S. S., Shamlou, P. A.: The Engineering Effects of Fluids Flow and Freely Suspended Biological Macro-Materials and Macromolecules. Vol. 67. p. 83

Zhong, J.-J.: Biochemical Engineering of the Production of Plant-Specific Secondary Metabolites by Cell Suspension Cultures. Vol. 72, p. 1
van Zyl, W. H. see Hahn-Hägerdal, B.: Vol. 73, p. 53

Subject Index

Activated charcoal 71
Acetyl-CoA 32, 42, 44
Acetyl-CoA carboxylase 44
Acetyl-CoA synthetase 44
Actin binding proteins 108
Actinorhodin 39, 41
Acyl carrier protein (ACP) 40–43
Acyl-CoA 32, 42
Aloesaponarin II 40
Amino acids 42, 46
Amphotericin 32
Amylase 113
Anaplerotic cycling 21
Aspartate kinase 22
Aspergillus 103
Aspergillus awamori 112
Aspergillus oryzae 113
Auxostat 148
Avermectin 32, 33, 46

Backfluxes 11, 16
Breeding 54, 73–74, 78
Butyryl-CoA 44–46
By-product formation 65

Candida albicans 120
Carbohydrates 42
Carboxylases 12
Cell banking 36
Cell wall 107
Cell wall permeability 11
Chemostat 76–77, 91
Chitin synthases 117
Cholesterol 32
Clone selection 35
CoA ligase 46
Compactin 32
Crotonyl-CoA reductase 46
Culture preservation 35
Current good manufacturing process (cGMP) 36
Cyclic fluxes 16

Daunorubicin 41
6-Deoxyerythronolide B 40
Deoxyerythronolide B synthase (DEBS) 47
Deoxymethynolide 40
Dihydrodipicolinate synthase 23
Dimorphic organisms 119
Dimorphism 103, 120
–, molecular level control 123
–, structural changes 122
Diploid 73
DNA array 5
DNA shuffling 131
Doxorubicin 32, 33
Drug delivery 3

Enzyme production 112
Epothilone 39, 40
Erythromycin 32–35, 39–42, 46, 47
Escherichia coli 39–42
Ethanol 54–79
Ethanol yield 61, 65, 68, 70, 72, 75
Ethylmalonyl-CoA 46
Expression increase 26
Expression level 71

Facilitated diffusion 62–64, 67
Fatty acids 42–46
Fed-batch process, control 38
Fermentation analytical tools 38
Fermentation inhibitor 57, 60, 61, 73
Fermentation, small-scale 35, 38
Filamentous fungi, morphogenesis 104
FK506 32, 33
FK520 46
Flow cytometry 151
Flux analysis 153
Fluxes, cyclic 16
–, in vivo 26
–, reverse 12
Frenolicin 39
Fungal morphogenesis 104
Fusion protein 66

GC-MS 6
Gene directed mutation 26
Genomics 47
–, functional 6
Glutamate production 11
Glycerol transport 67
Glycolysis 70, 73, 77
Granaticin 40

Hemicellulose 55, 57–61
High copy plasmid 41
Holo-ACP syntahse 42
Hxt1-7 deletion strain 64
Hyphal architecture 105
– –, engineering 116
Hyphal extension, apical 106

Immunosuppressant 32
Indan oxide 87–101
1,2-Indandiol 87–100
Indinavir sulfate 87
Inoculum procedures 38, 39
Isobutyryl-CoA 45
– mutase 45
Isotopomer labelling 12, 15

Lignocellulose 54–61, 73, 75
Limonene-1,2-epoxide hydrolase 99
Lovastatin 32
Lysine exporter 22
Lysine production 11, 26

Magnaporthe grisea 115
Malonyl-CoA 32, 42, 44
Malonyl-CoA decarboxylase 44
Malonyl-CoA synthetase 44
Media, defined 37
Media development 35–37
Metabolic control analysis 2
Metabolic engineering 1
Metabolic flux analysis (MFA) 76–77, 90, 94
Metabolite balancing 12
Methionine 11
Methoxy-malonyl-CoA 46
Methylmalonyl-CoA 32, 40, 42–46
Methylmalonyl-CoA decarboxylase 45
Methylmalonyl-CoA epimerase 45, 46
Methylmalonyl-CoA mutase 45, 46
6-Methylsalicylic acid 39, 40
Methymycin 39
Microcolonization 144
Microtubule associated proteins 108
Modeling 54, 71, 75
Monensin 32, 33, 46
Morphological engineering 103

Mucor rouxii 121
Mutagenesis 34, 35, 47, 134
–, site-directed 66
Mutator 137

NAD(H) 65–68, 76, 77
NADP(H) 65–68, 70, 72, 76, 77
Naphthalene dioxygenase 88
Narbonolide 40
Neurospora crassa 109
NMR 14, 17

Oleandomycin 39
Oligonucleotide array 5
Overpopulation 34, 39, 47
Oxygen 65, 68
Oxygen-limitation 61, 66, 70
Oxytetracycline 33, 39, 46

Parallel carboxylations 12
Penicillin production 111
Penicillium chrysogenum 111
Pentose phosphate pathway (PPP) 70–73, 76–78
Periodic selection 141
Phospholipases 120
Picromycin 39
PKS cellular protein levels 40
Plasmid instability 41
Plasmid stability 159
Polyketide precursors 32, 38–47
Polyketide synthase (PKS) 32, 34, 39–47
Polyploid 54, 57, 73
Polysaccharide 57–59
Post-translational modification 40–43
Process development 35, 36
Propionibacterium shermanii 46
Propionyl-CoA 32, 42, 45, 46
Propionyl-CoA carboxylase 42, 45
Propionyl-CoA synthetase 45
Protein kinases 114
Proton symport 62–64, 78
Pseudomonas putida F1 89

Radiocarbon labelling 13
Radiolabelled tracers 93
Random mutagenesis 54, 57, 70, 74, 75, 78
Rapamycin 32
Reverse fluxes 12
Rhodococcus sp. 88–91
Rifamycin 32, 33

Saccharomyces cerevisiae 115
Saccharopolyspora erythraea 35, 41, 42, 46, 47

Saprolegnia 108
SARP 41
Scale-down 39
Screening 151
Signal transduction 4
Site-directed mutagenesis 66
Spinosyn 32
Strain improvement 34-36, 47
Streptomyces antibiotic regulatory protein (SARP) 41
Streptomyces cinnamonensis 46
Streptomyces coelicolor 39-41, 47
Streptomyces collinus 46
Streptomyces lividans 39, 41
Streptomyces peucitius 41
Streptomyces promoters 41
Stress resistance 157, 158
Succinyl-CoA 42, 45, 46
Superhosts 46
Systems biology 7

Tetracenomycin 39
Thioesterase 42
Toluene dioxygenase 87, 89
Tracers, flux determination 94-98
–, radiolabelled 93
Tricarboxylic acid (TCA) cycle 77
Tylosin 32, 46

Valine catabolism 45

Xylan 55
Xylitol 56, 65-68, 70, 77, 78
Xylitol dehydrogenase 56, 65
Xylitol transport 67
Xylose 54-79
Xylose isomerase 55, 57, 65, 68-70
Xylose reductase 56, 65, 66
Xylose transport(er) 54, 57, 61-64, 78

Yeast 42

MIX
Papier aus verantwortungsvollen Quellen
Paper from responsible sources
FSC® C105338

If you have any concerns about our products,
you can contact us on
ProductSafety@springernature.com

In case Publisher is established outside the EU,
the EU authorized representative is:
Springer Nature Customer Service Center GmbH
Europaplatz 3, 69115 Heidelberg, Germany

Printed by Libri Plureos GmbH
in Hamburg, Germany